HOMELESS

Policies, strategies, and lives on the street

GERALD DALY

LONDON AND NEW YORK

First published 1996
by Routledge
11 New Fetter Lane, London EC4P 4EE

Simultaneously published in the USA and Canada
by Routledge
29 West 35th Street, New York, NY 10001

Routledge is an International Thomson Publishing Company

Typeset in Photina by Keystroke, Jacaranda Lodge, Wolverhampton
Printed and bound in Great Britain by
Biddles Ltd, Guildford and King's Lynn

British Library Cataloguing in Publication Data
A catalogue record for this book is available from the British Library

Library of Congress Cataloguing in Publication Data
Daly, Gerald P.
Homeless / Gerald Daly.
p. cm.
Includes bibliographical references and index.
1. Homeless persons. 2. Homelessness. 3. Homelessness–Government policy. I. Title.
HV4493.D35 1996
362.5′8′0973–dc20 95–26465

ISBN 0–415–12028–4 (hbk)
0–415–12029–2 (pbk)

HOMELESS

The causes of homelessness are disputed by both Right and Left. But few would argue that either massive government programs or regressive social cutbacks have succeeded in addressing the complexity of homelessness. More people are now in temporary shelter and on the streets than ever before.

Poverty, unemployment, deinstitutionalization, and economic dislocation are among the leading causes of homelessness. Voluntary organizations point to the failure of emergency shelters and food banks, the reductions in social programs, and the severe shortage of affordable housing. On the international scale, the changing global system has placed new demands on the economies of Europe and North America which have had a negative effect on resource allocation, employment, and even political will.

This book is the first comprehensive international study of homelessness. It points the way to new, constructive approaches to this critical social issue. The author demonstrates that the definition of "homeless" must itself be broadened, to encompass those at immediate risk of dispossession as well as those chronically without shelter, if homelessness is to be effectively confronted (before and after it happens) by public policy, voluntary organizations, and individuals themselves.

Gerald Daly is Professor in the Faculty of Environmental Studies at York University, Toronto. He has worked with housing agencies and non-profit groups and has published widely on housing, homelessness, and comparative planning.

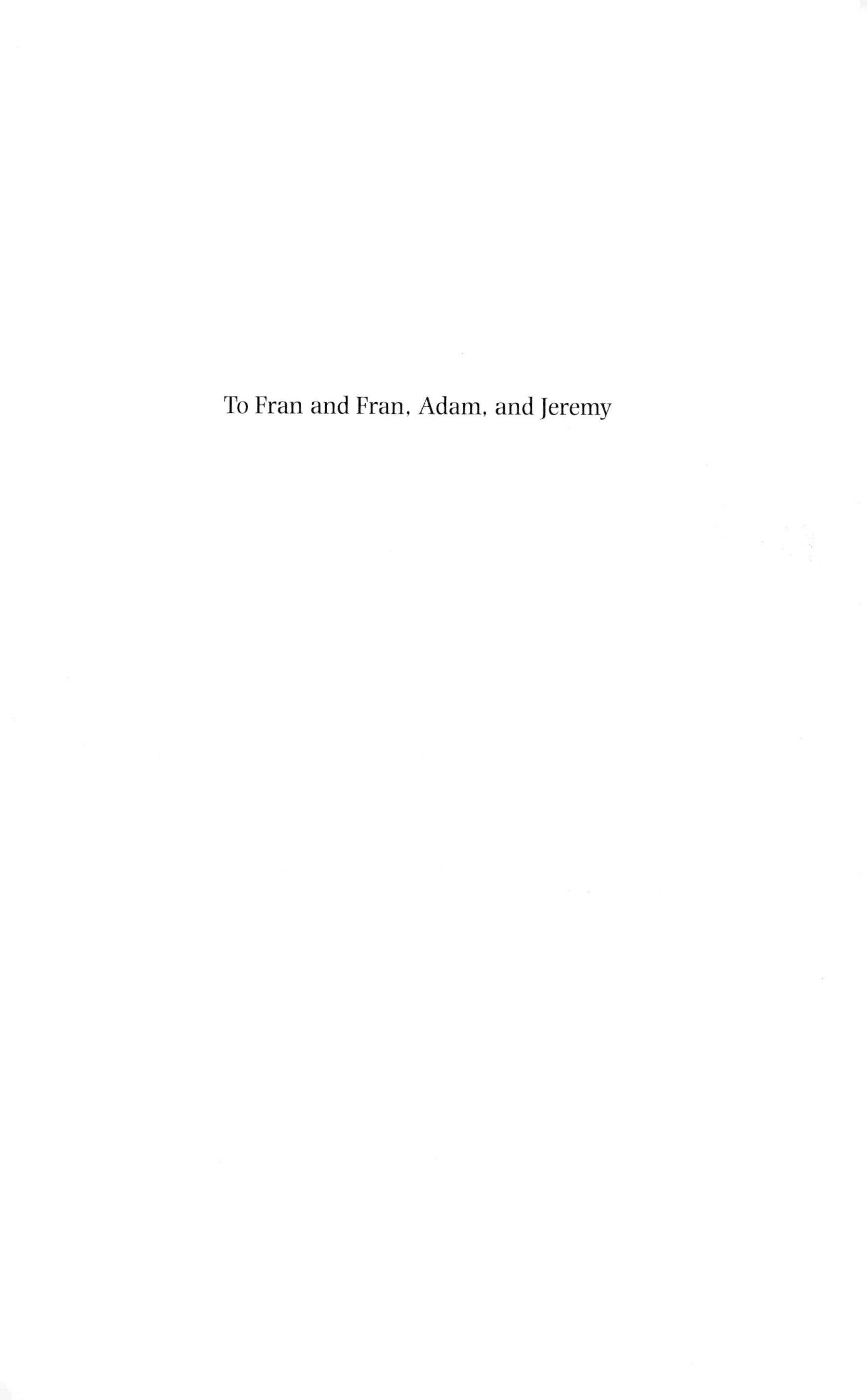

To Fran and Fran, Adam, and Jeremy

CONTENTS

PLATES

TABLES

PREFACE

During the 1980s, like other visitors to American cities and to the South Bank in London, I was profoundly disturbed by encounters with people who sleep in doorways or live in cardboard containers. These scenes provided tangible evidence of a public policy failure.

My concern about homelessness stems from its links with social policy and its pervasive nature. Affecting a wide variety of people, it is often the result of events beyond their control as individuals; at the same time, it is a problem which certain public and voluntary agencies and self-help groups have begun to address. I am interested in the nature, scope, and relative success of these responses to homelessness.

Homeless is a comparative exploration because I believe that it is possible to learn from the experience of other countries and from their attempts to design social policy in response to a common phenomenon. A study of Britain (within the context of Western Europe), the United States, and Canada is valuable because, while these countries encountered similar problems during the 1980s and 1990s, their reactions were different in many important respects. Understanding these varied responses to homelessness by public, private, and voluntary agencies is essential to assessing the advantages and shortcomings of various approaches.

While trying to avoid becoming enmeshed in circuitous numerical exercises, I have attempted broad estimates of the magnitude of homelessness. An analysis of the precipitating conditions and underlying problems is included, along with an exploration of the historical dimension of homelessness and changes in public policy since the late nineteenth century. The primary research focus of this book, however, is not on the "pathology" of street life or on the demographic characteristics of homeless individuals, as these have been documented by others (Bachrach 1983, Watson and Austerberry 1986, Burt and Cohen 1989, Niner 1989, Greve with Currie 1990, Golden 1992, Liebow 1993). Instead, I have concentrated on the institutional and policy contexts which give rise to homelessness and an analysis of the roles of public and private groups to determine who the key actors are, what they do, why they do it, and how they relate to one another. Considerable attention is given to specific issues associated with homelessness and to innovative measures designed to alleviate or preclude problems.

Several methods were used in this study, including literature review, analysis of policy debates and public programs, assessment of voluntary sector efforts, and interviews with a broad array of homeless people, policy makers, and care-givers. These approaches were selected to ensure broad coverage and to report on a number of different perspectives in order to illustrate the diversity of homeless people, the complexity of the issues, and the wide range of possible responses.

The first section of this book provides an overview of the issues, attempts to define the central concerns (focusing on poverty, housing, and the politics of distribution), and examines the evolution of homelessness in Britain, the United States, and Canada. Because Britain is now, perhaps reluctantly, a part of the European Community, students of housing policy and social issues in the U.K. must look to both Western Europe and North America for comparative data. Accordingly, in Chapter 4 I have briefly reviewed the situation in Western Europe in order to provide context and understanding for the British and North American case studies.

The second part of this study describes the human experience of homelessness, including health issues, deinstitutionalization, and the precarious nature of life on the streets and in emergency shelters in order to understand, in personal rather than statistical terms, who is affected and how.

Responses to homelessness are revealing of both government policies and public attitudes. The final section of this study examines the measures introduced by public and voluntary agencies and, in a few cases, those actions taken by private enterprise or by joint ventures involving several parties. Throughout this century a common reaction to homelessness has been inaction or a tendency to attribute problems to individual failures. Increasingly, however, many imaginative schemes are evolving as agencies develop a better understanding of the complexity of issues affecting homeless people as well as the diverse nature of their situations. Considerable attention is directed to these innovative responses because it is important to offer alternatives to traditional responses rather than simply deploring the magnitude and obdurate nature of homelessness.

I hope that *Homeless* will make a contribution by providing readers with substantive information and by demonstrating the potential of creative approaches, particularly those grass roots and self-help initiatives designed to assist people before they are forced to the streets.

ACKNOWLEDGEMENTS

In conducting this study I have relied on the assistance offered by numerous individuals in Britain, the United States, and Canada. Included are Margaret Blakeney, Barbara Brook, Carina Hernandez, Gary Fields, Alison Guyton, Pat MacBain, Ray Rogers, Janet Thompson, and Deborah Wandal. I am particularly indebted to Lous Heshusius for her incisive comments on several chapters.

Photographs were provided by Joyce Brown of Toronto, Shelter in London, Giroscope in Hull, England, the U.S. National Archives, the Greater London Council, the Library of Congress, Dover Publications in New York City, the City of Toronto Archives, Rod Hackney, the Pine Street Inn in Boston, Massachusetts, Byron Bignell and Streethealth in Toronto, as well as the Paul Sullivan Housing Trust in Boston. I received generous research assistance in the form of grants from the Social Sciences and Humanities Research Council of Canada, the Ontario Ministry of Housing, York University, and the Faculty of Environmental Studies.

Finally, I gratefully acknowledge all the individuals, whose names do not appear here, who generously gave of their time to discuss homelessness from a personal perspective.

PART I

UNDERSTANDING HOMELESSNESS: A CONTEXTUAL PERSPECTIVE

Why did homelessness emerge during the 1980s as a public policy dilemma? What can be done about it, and what are the implications for Western societies? *Homeless* is about the nature and causes of homelessness, the policies designed to cope with this growing problem, and the responses by governments and the voluntary sector in Britain,[1] the United States, and Canada. This review will focus on several key issues:

- the relationships between central and local governments
- the effectiveness of public policies
- gaps between policy and practice
- responses to homelessness and the nature of innovative projects created by public, voluntary, and self-help organizations.

Recent increases in homelessness are attributable to global economic changes, a severe shortage of affordable shelter for low-income households, and cutbacks in social programs. Among those at risk are single mothers, battered women and children, abused youths, disabled and frail elderly individuals, and the families of workers whose jobs have disappeared. The conventional response, from both the private and public sectors, was to provide emergency shelters and food banks. These expedients failed, however, to confront underlying problems of poverty, housing provision, employment, and resource allocation.

Because many people who are inadequately sheltered will lose their housing for at least a short time, homelessness must be broadly interpreted to include those at risk. As interpreted in policy and programs, definitions determine who receives assistance, the amount and type of aid provided, and by whom. If construed very narrowly it is easy to rationalize the problem as one peculiar to deviants. I have opted for a more liberal definition: people are considered homeless if they lack adequate shelter in which they are entitled to live safely. At the extreme, they are sleeping rough. Others live under a roof but their accommodation is lacking in safety, security, or basic amenities (e.g. heat, water, bathroom). Homelessness is a fluid and elusive concept. People who lack secure accommodation frequently change location, status, and living arrangements. Their deprivation depends on the extent to which the absence of shelter is combined with social isolation and economic poverty.

1

INTRODUCTION

> The second half of the twentieth century may someday be recalled as the time that we became painfully aware of the social and ecological costs of industrialization . . . We cannot rely on normal market forces nor on people's best intentions to save their environments and themselves . . . In the 1960s and 1970s . . . the only thinkable solution to commons dilemmas was government intervention. [Now] . . . the same problems and the same theory trigger discussion of another solution: privatization.
>
> (McCay and Acheson 1987: xiii–xiv)

Inevitably, government policy as well as the literature on homelessness is politically or ideologically motivated. Writers on the left evoke memories of the 1960s: severe social problems require massive government intervention and a great deal of money. Charles Murray and others on the right suggest that social problems are attributable to the existence of an "underclass of . . . poor people who chronically live off mainstream society (directly through welfare or indirectly through crime) without participating in it" (Murray 1984: 5).

Neither argument is persuasive. Both emanate from obdurate ideological stances. A more convincing case is made by Ellwood (1988) who proposes alternatives to welfare, including education, health care, training, and work incentives that do not stigmatize participants. Particularly in the United States, and to a growing extent in Canada and Britain as well, the resources and political will to significantly expand social spending do not exist, or will not be made available. Increased expenditures are unlikely unless they are demonstrably cost effective, socially useful, and politically acceptable.

In the past, efforts rooted in extreme ideological beliefs either were mean-spirited or they equated high levels of public spending with compassion, engendering resentment on one hand and learned dependency on the other. A more balanced approach is needed; one which emphasizes public and voluntary sector programs that are sensitive to actual needs while simultaneously fostering an individual's capacity for self-help.

A CONCEPTUAL FRAMEWORK

Acknowledging the historical transformation in the debate over public policy, how does one construct an approach which links shrinking public mandates and fiscal restraint, changing conceptions of "community," and increasingly unregulated global economic forces? Whereas the response to a range of social and economic problems prior to 1980 was seen in terms of government intervention, the world has become increasingly globalized. Notions of a "community sacred trust" or a sense of political commonwealth are undermined by privatization, deregulation, and free trade; but these concepts, an outgrowth of globalization, become harder to justify when one steps off the "level playing fields" constructed by capital and global markets. One question which might be raised, then, is the extent to which the issue of homelessness is the result of a loss of community (or the public realm) as a result of these forces?

The state and a civil society

Charles Taylor argues that society can function outside the political realm; that "society is not constituted by the state but limits it" (Taylor 1995: 287). One of the concepts underlying this book is that, while much of our behavior is atomistic, the self is not entirely unencumbered but is situated as a social being; therefore, we have social obligations. More than an aggregation of individuals pursuing benefits through common action that they cannot secure individually, a civil society is based on the notion of community. It is animated by a conviction of a shared common good. In this sense, I distinguish between the civil society and the state. The former has a web of autonomous associations independent of the state. Civic humanism connotes individual freedom. But freedom is not untrammelled; it is constrained by the social responsibility to ensure that public goods are distributed in such a way that everybody has enough to live at a minimally acceptable standard. There are, indeed, what Taylor calls "irreducibly social goods" (Taylor 1995: 192).

The state exists to provide for the common good. One of its principal functions is to ensure redistributive justice (through transfer or tax payments, for example), so that social (or public) goods are equitably allocated among members of society and the disparities between haves and have-nots are minimized. But the power of the state is limited. National governments must devolve responsibility (along with funding authorization) to the local level, where there is presumably a greater appreciation of local issues. The government must also ensure that generally acceptable minimum standards are met, because of the propensity of some local officials to ignore social problems.

Certain aspects of this function may be taken on by the voluntary sector (with government financial assistance). There is clearly a role for non-profit organizations with an understanding of local problems and the needs of people, with low-cost programs already in place, and with a minimum of unnecessary bureaucracy. Caution must be exercised, however, to ensure that this transfer of power does not become simply a guise for privatization, absolving public authorities of all responsibility. This issue is frequently occluded by those who pursue an uncompromising agenda of rugged individualism, laissez-faire capitalism, deregulation, and obsessive privatism: "the rhetoric of increased privatization . . . functions as the rationalizing agent of public unaccountability and, ultimately, irresponsibility" (Williams 1991: 47).

Globalization

A central unifying concept is the nature of global economic shifts and their reflection in political decisions to limit social spending. World economic activity quintupled during the final four decades of the twentieth century, aided by support from politicians. During the 1980s and 1990s economic growth was embraced by neo-conservatives, led by Britain's Thatcher and Major, by Reagan, Bush, Gingrich, and Dole in the United States, and by Brian Mulroney, Prime Minister of Canada from 1984 to 1993. Connections among their political ideologies, globalization, deregulation, and privatization are evident. The line demarcating the public (government) and private (business) sectors is obscured. Key decision-makers move readily from one sector to another. To a significant extent, then, public policy is shaped by private interests.

"Globalization is seen to be the intensification of global connectedness, the constituting of the world as one place" (King 1995: 220). It is a multi-dimensional process which is transforming the economic, political, cultural, and social geography. Globalization is characterized by the concentration of economic control in multinational firms and financial institutions, worldwide networks of production, exchange, communication and knowledge, trans-national capital, and a freer flow of labor, goods, services, and information (Castells 1991: 307–347). In many large cities it has ushered in a new era marked by increased immigration, high unemployment, the rise of a large service sector, and commodification or privatization of social services.

The operation of such places as London, New York, and Toronto depends on infrastructure which permits "global control capability" and includes a ready supply of low-paid service workers, mostly women, recent immigrants, and members of minority ethnic groups, whose presence in urban areas (both in terms of residence and workplace) engenders class polarization and new forms of spatial disparity (Sassen 1991: 282). Immigrants are highly concentrated in certain cities and in particular districts: over 35 per cent of New York City's population is foreign-born and the city receives one out of four immigrants to the United States; it has the highest concentration of West Indians; along with Los Angeles it has the highest concentration of Hispanics; with Los Angeles and San Francisco, it has the country's highest concentration of Asians. In Canada, the great majority of immigrants stay in Vancouver, Montreal, or Toronto: because of its attractiveness to Hong Kong flight capital, Vancouver now has the largest Chinese community outside Asia. In Britain 60 per cent of Asians and almost 75 per cent of Afro-Caribbeans are found in (different districts of) the four largest conurbations (Cross and Waldinger 1992: 158–159).

Globalization is characterized by tensions, evident in the highlighting of particular ethnic divisions, fragmentation (along racial lines, for instance), and in the reinforcement of cultural and social prejudices and boundaries (King 1995: 221). Global cities are spatially stratified; inner city neighbor-hoods are segregated along income, class, ethnic, and color lines. The enclaves where new arrivals live and work appear on the city's topographical map as pockets of poverty in close proximity to gentrified districts adopted by well-paid professionals. This "peripheralization of the center" is exemplified by the social ecology of East London, for instance, where new office towers in the Docklands sit cheek by jowl with the most deprived boroughs of the inner city. Segmentation is manifest in the built form. The city and suburbs are characterized by single-use spatially disaggregated zones, including areas that are judged to be central and others which are peripheral or inconsequential. Urban social processes help define urban space (and its occupants), organizing, controlling, and commodifying them to suit the needs of capital (or "the market").

Race, class, gender, disability, and place define one's quality of life in these new urban centers. Minorities and migrants discover that their fortunes are tied to the future of the city. But they also find that their labor may no longer be valued. Many are unable to fill new jobs except marginal, unprotected, part-time, or temporary service sector positions, and they are locked into inner city ghettos by market and exclusionary forces (Cross 1992: 16). These processes engender widening gaps between haves and have-nots, an increasing likelihood of social conflict, displacement, poverty for those left behind by labor market changes, and, for some, homelessness. Anger fomented by social conflict is not directed at economic elites, however, but at other groups in the ghetto. Los Angeles rioters in 1992 vented their frustrations against Korean grocers and entrepreneurs. Gunnar Myrdal's observation in *American Dilemma* (Myrdal 1944: 68) was prescient: "the lower class groups will to a great extent take care of keeping each other subdued, thus relieving to that extent the higher classes of this otherwise painful task necessary to the monopolization of its power and advantages."

As the economies of world cities have become international in scope, the transnational movement of capital and goods has been followed by migratory populations. Divisions of labor are apparent – between men and women, illegal and legal immigrants, ethnic groups, and new arrivals and natives. Different ethnic minorities have gravitated to particular niches based on their skills, networks, and resources. Many have adapted by relying on the informal economy which is outside the regulatory apparatus of land use and building codes, health and safety requirements, workers' compensation legislation, and minimum wage laws. Others have drifted into illicit activities such as drug dealing. A great number, though, serve as replacement labor, filling openings generated by the exodus of whites. Other jobs for migrants have resulted from whites *returning* to the city: gentrifiers create a market for personal services and domestic labor. These are generally poorly paid positions; workers are typically part-time or casual employees who lack benefits and are not protected by labor legislation.

For those in the lowest income segments, the benefits of growth have not trickled down and income inequalities have intensified. Despite prosperity for those at the top of the income pyramid, roughly one quarter of the population of such global cities as London and New York are below the poverty line. Poverty and homelessness intersect with age, gender, and race, disproportionately affecting children, single mothers, and non-whites. As poverty has become identified with particular social groups, others in society have withdrawn from the public realm, closing the door on the notion of the city as commons.

The deserving and undeserving poor

Attempts to define homelessness and estimate its magnitude represent more than an academic exercise. Definitions embody political statements as well as value judgments. When reference is made to the dichotomy between the deserving and undeserving poor, for instance, an argument usually follows for increased social control and stringent limitations on government intervention. For those who are seen as undeserving (or on the periphery), the public policy response has not changed substantially over the past century. Warehouse-like emergency shelters, evoking memories of poorhouses or police station lodgings at the end of the nineteenth century, remain the principal means of providing for transients and chronically homeless individuals. Definitions of homelessness are revealing of vested interests. Some critics argue, for example, that public policy and voluntary sector activities are designed not so much to help homeless persons as to maintain existing institutional arrangements and to forestall radical change. "Compassion," Peter Marin believes, "is little more than the passion for control" (Marin 1987: 48).

Definitions reflect different purposes, values, ideologies, and political agendas. Some government agencies use narrow demarcations, ignoring people who are not on the streets or in emergency shelters (U.S. Department of Housing and Urban Development [HUD] 1984: 19). Responding to widespread criticism for the exclusion of people in welfare hotels and insecure living arrangements, HUD's 1984 definition (which was primarily limited to shelters) was later enlarged by Congress (U.S. Congress 1987) to include people in welfare hotels and temporary or transitional housing.

Others argue that when responses concentrate on the vulnerability and helplessness of the "client population," the potential political power of this group is ignored while their dependence on the existing social service system is encouraged. Some observers, notably Dear and Wolch, view homelessness – which they term the most extreme consequence of deinstitutionalization – as part of a network of social and economic problems, attributable in part to the actions of governments and service providers (Dear and Wolch 1987).

Considerable disagreement exists among researchers about the nature of homeless people, the causes of their condition, and their capacity for remedial action. This is an important issue because, in many instances, the descriptions used become labels which can pigeonhole individuals in negative ways. Once classified as "the underclass," for instance, those who are down and out are seen by many as hopeless, a latter-day version of the "undeserving poor." Categorization may also imply extreme degrees of isolation or disaffiliation which have implications for social programs, given the current political popularity of cutbacks in social spending.

During the twentieth century, descriptions of homeless people have ranged from sociological surveys of "tramps and deviants" on Skid Row, to sympathetic portraits of destitute families during the Depression, to reports by advocates in the early 1980s designed to dramatize the plight of street people. In the mid-1980s the magnitude of homelessness and the nature of emergency shelter responses dominated discussion. Debates over the accuracy of population counts ensued, while public agencies and advocate groups remained poles apart. Gradually, the notions of vulnerability and "shelter uncertainty" gained acceptance and studies began to take account of "the invisible homeless." After mid-decade researchers examined the characteristics of homeless people. Subsequently, a few studies emerged which addressed causes – focusing on poverty, the lack of affordable housing, cutbacks in social services, and deinstitutionalization.

Researchers are now beginning to address the way society categorizes and responds to homeless individuals. Rather than talk about homeless people as being "disaffiliated," some suggest that it is more appropriate to examine the notion of dissociation as it is applied to homeless people, victims of disasters, and poor inner city residents. These individuals are consigned to the periphery of public consciousness because by failing to conform they violate social norms and offend public sensibilities. We deal with them by dissociation, distancing ourselves to minimize or displace feelings of resentment, fear, contempt, guilt, shame, or conflict. In doing so a cycle of disinterest and disaffection is generated, allowing us to shun collective responsibility. We compartmentalize and place barriers between "us" and "them." We tend to see some things and to ignore others. As a dehumanizing process, the extent of this dualistic dissociation is manifest in the terminology used to describe homeless individuals. Common use of the term "the homeless" instead of "homeless persons" or "people without houses" facilitates the distancing process. "They" become an amorphous, remote, alien mass lacking individuality or even humanity. A sense of community is lost. Definitions and descriptions of "the homeless" expose our personal values and beliefs, especially when homelessness is characterized by what it is not. Our egos yearn "to be free of complexity and of change, of relation, and of needing to know the irksome other" (Keller 1986: 174).

In linking the irksome other and the "peripheralization of the center," a connection can be discerned between homeless people and other so-called undesirables in terms of the dominant society's exclusionary behavior. In working with squatter communities in Nairobi I have found that there are almost 100 "informal settlements," some of which are shown on official maps as "forest." If such shantytowns are not recognized by public authorities then the municipality is not obliged to provide infrastructure or municipal services. One of the largest, Mathare Valley, has over 200,000 permanent residents and has existed for well over thirty years, but is not yet officially recognized and still lacks municipal services and infrastructure (Daly and Muirhead 1993).

Lifton suggests that our society engages in destructive patterns of symbolizing homelessness (Lifton 1992: 133). In Western societies we mystify homelessness. The inability to separate results from causes leads to mystification. By creating a form of analysis which hides the cause of problems, this process re-entrenches the status quo; it diverts analysis from underlying structures. As soon as the problem is mystified, it can be denied or dismissed as unwieldy, abstract or diffuse, even intractable:

> Despite the visibility of homelessness, extensive media coverage, intensive advocacy efforts, widespread voluntarism, government programs, and our familiarity as a people with this recurrent social problem, homelessness persists and an industry has grown up around it. How do you explain all this expenditure of energy with virtually no social transformation and none on the horizon?
>
> (Giamo and Grunberg 1992: 150–151)

The use and limitations of language

Another thread woven throughout discourse on homelessness is the use of language and the ways in which words, concepts, values, and beliefs shape our behavior and our views of others. By exercising our power to name, we construct a social phenomenon, homelessness, the criteria used to define it, and a stereotype of the people to whom it refers.

Language, as used here, is broadly construed. It includes media images, sound bites, and defamatory rhetoric, as well as policies and programs which convey mainstream society's messages of power, influence, and authority. These messages, which raise a number of ethical dilemmas, can become tools of manipulation. Homeless individuals may be silenced by such power relationships, control mechanisms, and by messages contained in popular media. Language is instrumental in how we construct reality. Certain keywords have meanings, express values, and proclaim ideological positions such that they are inextricably linked with the problems which they are used to discuss. Language is political. It is revealing of how we look at social, economic, and political issues; and the ways we use language serve to define relations of power.

To the extent that it is visible and audible, the agenda of the political debate is set by public officials and experts (academics, consultants, and representatives of voluntary agencies), while typically excluding the presumed beneficiaries. Because they lack a collective voice and are not organized, individuals on the street are represented by proxies whose interests may be self-serving. These relationships may constitute a control system based on a charity model and on naive assumptions about the need to dictate terms to the recipient population. A self-perpetuating network, characterized by common interests, mutual dependencies and benefits, it has fashioned a web of interdependent communities based on self-interest. It includes government agencies and bureaucrats, not-for-profit and voluntary organizations, professional care-givers and shelter operators. While most are well-intentioned they, nevertheless, are motivated by a desire to exercise power and a need for control: the power of the purse strings, the ability to set policy, to allocate resources, to plan and design programs, to decide who will be helped and who will not, to determine whose interests will be represented, and to sanction or condemn certain practices, values, or beliefs.

A separate, but related sphere includes the business community and the larger public, whose views on homelessness and charity help to define or constrain the roles of government and the voluntary sector and to determine the nature of opportunity/exploitation affecting the prospects of homeless individuals. Economic interests employ the language of competitiveness, laissez-faire, and the free market to describe their preference for a "level playing field." This field, however, is readily accessible only to members of the team. For others, who do not enjoy the advantages and resources of team members, the field is strewn with barriers and obstacles. The participants in this debate represent two dissimilar cultures, based on unequal power relationships with discordant values, beliefs, and languages. By describing how their worlds intersect I will try to clarify the nature of the bureaucratic and institutional contexts in which policy is formed and the arenas in which political and power dynamics play themselves out. Context is centrally important. It shapes the terms and tone of the debate over homelessness and helps to define the issues. It determines what gets discussed in the public domain as well as what gets suppressed or excluded from the agenda.

In social, economic, and political discourse the life stories of homeless people are typically devalued, shunted aside, or unconsciously limited. Spurious forms of domination, which prejudice the prevailing view of this "alien" culture, raise issues with respect to the social location of knowledge. Whose stories are valued or not valued? What is considered real, significant, or valuable information? Through this book I will try to explain the political, economic, and social contexts affecting homeless people. Everyone has the right to "name the world," to describe their lives, to explain the human condition from their own perspective. Paul Shepard tells us that "there is a secret person undamaged in every individual" (1982: 129). To hear and understand them, however, requires that we listen. We need to understand their situation in terms which emancipate rather than relegating them to invisible areas where they do not intrude on the public conscience.

An alternate reality

"Our cultural environment plays an active role in the questions we ask" (Prigogine and Stengers 1984: 307). This should cause us to re-examine some of our underlying assumptions which guide research into homelessness. Straightforward notions of causality, for example, and linear relationships among variables affecting homelessness are not necessarily appropriate. The links among poverty, economic change, deinstitutionalization, demographic shifts, employment, marginalization, and the worlds of homeless people must be clarified. Such complex interacting processes of turbulence and social change are not readily reducible to generalizations or linear explanations. The relationships among them vary according to context. This concern for context and for situated knowledge explains, in part, the motivation for cross-national studies.

Traditional concepts of poverty and homelessness rely on mechanistic models which describe a closed, relatively stable, orderly system in equilibrium.

Prigogine and Stengers offer an alternative vision of open, dynamic social systems which is illustrative of the current period of rapid economic and social change; a time characterized by "disorder, instability, diversity, disequilibrium, nonlinear relationships (in which small inputs can trigger massive consequences) and temporality – a heightened sensitivity to the flows of time" (Toffler 1984: xiv). This suggests that attempts to understand complex, dynamic, pluralistic systems in dualistic and deterministic terms will fail.

The survival strategies of homeless people may be regarded as spontaneous, random, contradictory, illogical, ill-conceived, or even bizarre. But judgments of this sort, exposing the contents of mainstream society's cultural baggage and the need to hold the irksome other at arm's length, fail to comprehend the imperative of adaptation and change in a world which does not conform to dominant conceptions of "reality." Though the universe of street people is different, it typically is not aberrant. Life on the streets requires adherence to a different set of organizing principles, an alternate reality. Take the notion of time, for example. Anthropologists accept that different cultures have very distinctive conceptions of time. This idea may be applied to homeless people as well. To the academic researcher, linear time may be a central, defining, organizing notion. To a street person, however, the truism "time is precious" – one of the mystifications of capitalism – may be laughable. Time is plentiful. What we value, they say, isn't time but the necessities of life, a sense of security, and hope for the future.

For persons without shelter the line between chance and necessity may be hazy. Exogenous change is inherent in their "system." In our lives most of us believe that we seek rationality, coherence, and predictability. People on the street, however, try to find immediate comfort, short-term satisfaction, and medium-term survival. Past, present, and future are blurred. In an opaque, uncertain, turbulent, fatalistic world it is difficult to distinguish between preparing for the future and living for the present. Often people on the street appear to be resigned to their fate: "I'll always be poor. After all, I come from a broken family and I'm a fourth-generation welfare mother" (a 35-year-old single mother in Toronto).

The environment of homeless people, then, while apparently characterized by disorder and chaos, has an order of its own. To detached observers their lives may seem schizophrenic – for instance, attempting to satisfy basic needs while engaging in what appears to be self-destructive behavior. Yet these seemingly incompatible strands may connect in some coherent fashion. Frequently, individuals on the street lack the power to control events which affect their lives. Chance plays a major part. The "system" within which they operate is not in equilibrium. They are powerfully influenced by external forces. Seemingly small changes can become amplified, often with devastating results. When researchers describe this world they need to define "reality" in situational terms. A more open, contextual, and dynamic conceptualization of street reality – situated knowledge – is needed; one that is grounded in an attempt to understand the sense of order and the coping strategies created by homeless persons; one that reveals how they manage to live with isolation, anxiety, frustration, and rejection.

The role of the individual

The concept of "social death" is occasionally employed as a way of explaining homelessness in terms of isolation, invisibility, powerlessness, and alienation. Giamo refers to "this whole feeling of death, both the physical threat of death and kind of a lingering social death and a sense of disillusionment and despair" (Giamo and Grunberg 1992: 138). Lifton (1992) believes that homeless people are victimized by being treated as outcasts. Society relegates them, symbolically at least, to the living dead.

As a conceptual framework, however, this approach may paint with too broad a brush, obscuring individuality and essential differences. My research leads to a belief that homeless individuals yearn for a voice and for an integral role in determining the shape of their own futures. Virtually all have some capacity for helping themselves and, in many instances, they have successfully wielded political power, contesting urban spaces, manipulating the press and politicians to achieve their objectives. Some, too, will not be content to merely "fit in" with mainstream society. They may challenge the dominant structures, the status quo, and doing business as usual.

I am sometimes pressed to explain the role of individual responsibility in homelessness. To suggest that, in some cases, individuals bear partial responsibility for their homeless state – and this is accepted by most homeless people – is not to argue that being on the streets is the result of moral laxity or personality defects. It is not useful to engage in this exercise if it simply leads to "blaming the victim." I concede some degree of individual responsibility in certain cases because it may help to understand precipitating factors; it may be useful as well in emphasizing the role which can be played by homeless people in helping themselves. I also emphasize, however, that the causes of homelessness typically are systemic and dynamic rather than personal.

Isn't it true, some critics ask, that alcohol abuse may lead to the loss of employment, family, and housing? I agree that it may. It is also true that people who are poor and drink are far more likely to lose their jobs and houses than people who are well-off and drink, but who apparently are above reproach and immune from privation. An individual who is poor and resorts to alcohol to dull the pain undoubtedly makes things worse; but we cannot ignore the relationship between poverty and homelessness which pushed this person over the brink. Moreover, the link between homelessness and mental illness or clinical depression, which may lead to alcoholism, cannot be brushed aside. People affected by depression are hobbled by self-doubt or self-contempt. A range of experiences can trigger deep depression, from incest to racial discrimination, from loss of loved ones to lay-offs or eviction, from wife battering to loss of power in one's career. Catastrophic events and personal crises lead some to take refuge in the bottle, and inevitably increase the likelihood of homelessness. It may be misleading, however, to attribute this state solely to "individual responsibility."

Why then, despite numerous disadvantages, are women less likely to become alcoholic and apparently more successful at avoiding homelessness? The incidence of alcoholism among homeless men is three times higher than among homeless women. For physiological, emotional, and social reasons, described in Chapter 5, women drink less than men. Physiologically, they are less tolerant of alcohol. Women, generally, are introduced to alcohol later in life than men and society condones male drinking more readily than among women drinkers. Women are more likely to have care-giving responsibilities and to have more intimate attachments than do men, whether to children, to family, or to peers, which means they are more likely to have access to housing (often by doubling up) and less prone to live on the streets. Further, women and men are socialized differently in terms of their place within society. Males are often told, in a number of ways, that their task in life is to "make it on your own," whereas women are socialized to ask for help and to offer assistance. Again any reference to individual responsibility can only be justifiably made with respect, by taking into account the differential female and male socialization patterns regarding an individual's location in relation to other and to community. Because they are subject to violence and abuse many women go to great lengths to avoid literal homelessness. In some cases they are forced to remain in abusive relationships, against their will, because the only alternative, life on the streets, is worse.

Research issues

The self–other dilemma illustrates some of the issues facing researchers, not the least of which concern ethical questions. In examining homelessness it is difficult for researchers to place themselves in another's shoes. Studies of homeless people may be intrusive. Inevitably, they are value-laden. Surveyors approach their "subjects" encumbered by preconceptions and stereotypes. Relationships between individuals are affected by power and status differences which stand in the way of communication and understanding. What we observe is not nature itself, Werner Heisenberg (1970) reminds us, but nature exposed to our method of questioning.

People on the street have been scrutinized by researchers for some time. Dozens of descriptive and colorful books have appeared in recent years; but many fail to go beyond the immediate words of homeless people to examine underlying problems and dynamics. There is a danger as well in relying solely on the voices of advocates and care-givers who relate the problems and experiences of homeless people through their own personal and professional filters. Accordingly, I have consciously avoided lengthy citations from homeless individuals or care-givers as a primary source of information. Instead, I have used their eloquent words sparingly but, I hope, effectively and have provided a number of profiles to convey a sense of their characteristics and experiences.

This work, then, is shaped by the fact that my relationship with homeless individuals and professional care-givers, while not consciously disconnected, is essentially at arm's length. I cannot claim, as I believe few researchers can, to have shared their experiences or perspectives or to completely understand the difficulties encountered by both groups (care-givers and people without shelter) in coping with a disinterested society.

Cutting across class, ethnic, age, and gender lines, homelessness affects a broad range of social groups. It does, however, disproportionately affect visible minorities, single mothers, and those in the lowest socio-economic groups. Inextricably linked to poverty – which reached a twenty-seven-year high in 1991 – homelessness worsened measurably in the 1980s and early 1990s.

Homelessness is not a new phenomenon, nor are the responses it generates. Common reactions are to label and stigmatize people without shelter and to differentiate between the deserving and undeserving poor. These notions, reflecting an institutional response of social control, engender public passivity. But the issues associated with homelessness relate more to a lack of public resource commitment than to individual failures or behavioral disorders. Government actions, at both national and local levels, contribute to this problem. Homelessness is a product, in part, of public policy, and thus cannot be separated from its social, political, economic, and institutional context. A grave shortage of affordable housing, affecting even those low-income people who work full-time, has been aggravated by deindustrialization and global economic changes.

Homelessness may be seen as a manifestation of a loss of shared common ground or abandonment of the notion of the public realm in a civil society. Government funding is essential in order to ensure that adequate housing and related services are available for low-income people. But communities must be created by those who will live in them. Self-help housing, then, is given considerable emphasis in this study because it represents far more than shelter; it is an attempt to reclaim the commons and to develop a sense of community, a sense of place, a place to call home.

It is illusory to expect private charities to assume major responsibility for provision of social services. There is, of course, a role for the third sector. But privatization and devolution of responsibility to voluntary agencies has degenerated, in many cases, into an attempt by government to opt out of its welfare role.

Homelessness represents a continuum from people at risk to those who are without shelter on a temporary or episodic basis, to individuals who are absolutely or chronically homeless. Different homeless people have different needs. A continuum of needs can be matched with support services to help people, most of whom can achieve some measure of independence. A range of programs is needed to address the multiple problems and complex issues associated with homelessness. But even a full array of initiatives will not suffice unless action is taken early; timely responses are most important to preclude homelessness or lessen its duration and severity.

CROSS-NATIONAL COMPARISONS

The fact that there are common problems among nations does not necessarily lead to common solutions. Comparative studies may conclude that transfers of housing and planning concepts from one country to another fail to generate intended benefits. But policies adopted in one country cannot be fully understood in isolation. It is essential to explore social, political, cultural, and economic contexts before policies and programs can be interpreted accurately.

Cross-national comparisons with respect to homelessness are difficult because of differences in definitions, the availability and comparability of data, regional disparities, and substantial variance among governments in the degree of diligence exercised in housing low-income applicants. In comparing countries and viewing issues at the national level we may overlook or obscure differences, or even contradictory trends, at the regional or local levels. Nevertheless, comparative research permits us to look more critically at our own system, with the benefit of different perspectives (Williams 1986: 140).

A number of significant similarities are evident in Britain, the United States, and Canada, especially with respect to the impact of globalization. These countries are linked by broad social and economic trends, even if their responses to homelessness are different, which suggests that there may be a commonality in political pressure from the top, in pursuance of neo-conservative agendas. During the 1980s and early 1990s, the political leaders, Thatcher, Major, Reagan, Bush, and Mulroney (as well as some of their counterparts in Western Europe) saw themselves as having common cause in seeking to cut social spending and to rein in the expansion of the welfare state.

Many of the problems encountered by homeless people in the United States also are found in Britain and Canada. Because of differences in the composition of laws and social safety nets, however, the profile of people classified as homeless is different. The definitions of homelessness applied in Britain and Canada (as well as in parts of Western Europe) are broader than those used in the United States. The concept of vulnerability to homelessness has been more widely adopted in Britain and Canada, so that people are accepted as "priority need cases" who would not receive public assistance in the American system.

I have, nevertheless, attempted to make comparisons and to quantify the extent of homelessness in the three countries. My research indicates that about seven persons per thousand in Britain are accepted as being homeless by local authorities – though in central London the figure is much higher, about twenty-five persons per thousand. The annual total for 1992 was over 184,000 households (about 500,000 people). This is approximately four times higher than the number accepted in 1977, when the Homeless Persons Act took effect. Almost one-third of the country's homeless people are located in Greater London. It appears that the rise in homelessness halted after 1992; there was a slight decrease thereafter (6 per cent) in the numbers of households accepted as homeless (which may be attributable to stricter interpretation of qualification criteria), a somewhat greater decrease (14 per cent) in the number of homeless households in temporary accommodation, but an increase in the proportion (9 per cent) being housed in bed and breakfasts (Department of Environment 1994a).

In Canada approximately five persons per thousand population use the emergency shelter system. A nationwide survey of these shelters was conducted on January 22, 1987 by the Canadian Council for Social Development (CCSD). This census found that over 100,000 different people used the 472 facilities at some time during the year. Each year, during the late 1980s and 1990s, more than 26,000 different individuals used Toronto's shelter system, which is funded by the metropolitan government. Extrapolating from these census figures, estimates of Canada's homeless population range from 130,000–260,000 (Canadian Council for Social Development 1987: 1). This is only an order of magnitude estimate, however, because no one knows what proportion of homeless people opt for sleeping rough rather than acceding to the regimen of warehouse-like shelters.

In the United States, according to HUD and to more recent studies in a number of states, about 1.5–2.5 persons per thousand population are judged to be absolutely or temporarily homeless (i.e. users of the emergency shelter system). In 1984 HUD concluded that about 250,000 to 350,000 people were homeless. Although representatives of advocacy groups claim that the national count is between two to three million, it appears that these estimates may be inflated. Subsequent studies suggest that, with annual increases of 10–20 per cent in the shelter population through the late 1980s and early 1990s, the national total of homeless people in the mid-1990s was perhaps 500,000–600,000 (U.S. Department of Housing and Urban Development 1984; Freeman and Hall 1986; Burt and Cohen 1989; Momeni 1990; Burt 1992). Using data from 147 cities with 1986 populations over 100,000, Burt found a significantly higher incidence of homelessness – 1.8 per thousand, triple the rate in 1981 – in cities with both high poverty rates and high per capita incomes (which is consistent with the notion of bifurcated or divided cities discussed by Sassen and Fainstein *et al.*). These rates, based on the number of people using soup kitchens and emergency shelters, reflect a circumscribed concept of vulnerability, however, and appear to understate the extent of homelessness. Burt, for example, did not examine rural areas and estimated that rural homelessness is only one-third the level of urban areas; many other studies of rural regions, however, suggest that the number of people without adequate shelter is quite high, though homelessness may take different forms, may be less visible than in major cities, and is difficult to assess; respondents are not forthcoming and many who are vulnerable to homelessness move frequently.

A nationwide random sample (Link *et al.* 1994: 1907–1912) of lifetime and five-year prevalence of homelessness (including doubling-up) found higher levels of homelessness and vulnerability than are usually reported by point prevalence surveys: an annual estimate, derived from these findings, puts the national average at about 6–8 per thousand. Another study of Vietnam-era male veterans found that 8.4 per cent reported having had "no regular place to live for at least a month or so" (Rosenheck and Fontana 1994: 421–427). A U.S. government (Fannie Mae) working paper cited emergency shelter admissions: 2.8 per cent of the population of Philadelphia made use of shelters over a three-year period and 3.3 per cent of New Yorkers used shelters over a five-year period (Culhane *et al.* 1993). After sorting through these surveys I concluded that a conservative "order of magnitude" average of 3–6 literally homeless people (per thousand population as noted in Table 2), can be supported.

Precise numbers are not available, but evidence from a number of sources indicates that the extent of homelessness grew throughout the 1980s at a rapid rate in all three countries, leveled off in Britain after 1992, but continues to grow, albeit at a reduced rate, in both the United States and Canada. The U.S. Conference of Mayors conducts periodic polls of municipal governments to determine if demand for shelters and services is growing; their findings point to continuing increases, but at a slower rate than in the 1980s. Their figures are not necessarily reliable, however, as city officials have a vested interest in exaggerating demand so that they will receive federal funds. Caution must also be exercised because increased "demand" in some cases is largely a reflection of increased supply; usage data rose rapidly in a number of instances because additional shelters were opened.

Credible statistics are available, however, for particular cities in North America, which have been compiled over several years. Data from Toronto, for instance, show that "the use of hostel services reached a high point in late 1987, after almost a decade of steady growth. A short period of decline ended in 1990 as demand once again increased" (Municipality of Metropolitan Toronto 1993). Rising demand continued through the first half of the 1990s, with shelter use growing by 10 per cent between early 1994 and early 1995. This growth is not across the board, though. While the number of women fleeing violent partners has risen rapidly, almost matched by a one-third increase in the number of families seeking shelter as a result of prolonged economic recession, there has been a continuous decline in the numbers of single adult men admitted during the 1990s. There are fewer homeless men because fewer jobs are available. Many young men left the city (as construction work virtually disappeared), housing supply expanded as rents for rooms and single room occupancy hotels (SROs) dropped, and potential migrants to the city never made the journey because job prospects were dismal (Municipality of Metropolitan Toronto 1995).

Differences in the numbers cited and the approaches taken in the three countries are due in part to different economic conditions (Britain's unemployment rate is substantially higher than the level in the United States, for example); to different perceptions of need (Canada's poverty line, in real terms, is more than 1.5 times as high as the figure used in the United States);[2] and to varying interpretations of homelessness. The definitions used by governments and, to some extent, by voluntary agencies, to determine eligibility for aid tend to be broader in Britain and Canada than in the United States.

Differences are also a function of the role of government and the importance accorded to housing issues. Housing is a fixture of Britain's political landscape. Throughout the twentieth century government housing activities have been a major factor in people's lives. Shelter issues have a political significance in Britain which may be hard for Americans to understand. It may also be difficult for North Americans to appreciate the enormous role played by local authorities (i.e. city councils and county councils), still a pervasive force in urban Britain. This helps to explain why a major plank in the Conservative agenda for Britain in recent years has been to privatize housing – 1.5 million council dwellings were sold prior to 1995 – to allow tenant management, and to pursue other strategies aimed at reducing government's housing responsibilities and weakening the role of municipal councils, many of them dominated by Labour.

A comparison of these countries, then, helps to develop a perspective on complex issues. While similar in some respects they have different traditions, governmental systems, and expectations regarding the role of the public and private sectors in housing and welfare issues. Not surprisingly, they have followed different paths in dealing with homelessness. Table 1 provides a general framework for comparing these approaches. This matrix suggests that several contextual dimensions shape responses to homelessness in these three countries. Prominent factors include the nature of central government and the relationships between national and local authorities; the extent to which collectivism[3] or individualism influence social policy; the effectiveness of social safety nets or the degree to which assistance is constrained by cost containment campaigns; the status of public health systems and the nature of care and accessibility for very poor or homeless individuals; and the current state of the economy in light of globalization.

TYPES OF HOMELESSNESS

Homeless people may be described in a variety of ways (e.g. demographic and familial characteristics, shelter history, health problems, degree of disability, work record, residence, etc.). Often homelessness is characterized in terms of presumed causes or precipitating problems: *accidental* (resulting from natural disasters or exogenous events); *structural* (relating to poverty or health problems, for example); *economic* (unemployment as a result of deindustrialization); *political* (refugees from areas of political or ethnic conflicts); or *social* (single mothers and others who are marginalized or discriminated against).

People without houses also are described in terms of the duration of homelessness or their degree of vulnerability, ranging from: those who are absolutely or chronically homeless; individuals who are periodically or episodically without shelter (e.g. migrant workers, young people, or women fleeing domestic abuse); those who are temporarily homeless as a result of an extraordinary event (sudden unemployment or severe health problem, death of household head, loss of one's home); to individuals who are vulnerable or "at risk" (e.g. single mothers with young children who are doubled up, frail elderly people, refugees, roomers, and lodgers unprotected by law).

Individuals lacking shelter also may be described along a continuum on the basis of their needs: (a) people who are at risk or vulnerable to homelessness in the near future, perhaps within the next month; they need short-term assistance in order to avoid being on the street; (b) individuals whose primary or sole need is for housing. Often these are working poor people who are temporarily or episodically homeless. They may require financial and other assistance to re-establish themselves but do not have other serious problems, provided they receive timely aid; (c) persons who can become quasi-independent but need housing as well as other supports such as literacy or employment training, budgeting and life skills assistance to enable them to manage on their own; (d) those who have substantial (and, often, multiple)

Table 1 Contextual framework for comparing the three countries

	Britain	*United States*	*Canada*
Government	Centralized; no constitution; legislature and executive very close; voting along party lines; local authorities have significant housing powers, but seriously weakened since 1979	Federal system; concentrated government power is mistrusted; laissez-faire and home rule important. Politics heavily influenced by race and parochialism	Federal system; significant role for the provinces; necessary to effect compromises among provinces and between provinces and the federal government in order to make major changes
Approach to social problems	Collectivism evolved over 100 years and is deeply entrenched; recent trend toward privatization	Individualistic; collectivism suspect; self-help important; some welfare benefits being privatized; no socialist party	Leans to British approach; socialism and protection of group rights more highly regarded than in U.S.
Status of social safety net In all three countries cost containment is a major issue	Established system of dole and hostels which works to maintain people but only at very minimal level. Program eligibility not as restricted as in the U.S.	Concern about long-term dependency on welfare; AFDC has grown dramatically; heavy reliance on private charity; recent workfare initiatives widespread	Function shared by local government (with senior government funding) and voluntary sector; government still assumes voluntary sector will carry major part of the load
Health In all three countries health problems are surprisingly pervasive. Few have publicly acknowledged the inequalities of the system	NHS still operating; cost problems; care generally good once people are admitted; tiered system: middle and upper classes opt for private care	Those who can afford it get good care. Those unprotected by employer or by private insurance (a third of the population) are vulnerable to catastrophe; may lead to homelessness	Good national system. Regional disparities; some provinces exploring privatization. Difficult for some homeless people, Natives, and rural residents to get good care

cont

Table 1 cont.	*Britain*	*United States*	*Canada*
Higher Education Basic problem in the three countries: making the system accessible to all and relevant to the new century	Still a class-ridden system; relatively few people in working class get higher education; as result there is relatively little upward mobility	Widely available though evolving into a two-tier system of public and private; higher education the principal source of individual social mobility	Widely available; relatively inexpensive university and polytechnic system and distance education. Government cutbacks affecting universities
Economy/ employment Globalization affecting all three countries; lack of imaginative schemes for job training and skills development especially in preparing for the 21st century	One of the U.K.'s major problems; unemployment over 2.5 million; some on dole for years; difficulty in maintaining productivity and competitiveness; concern about U.K. being left behind	Globalization and deindustrialization; low-paid employees marginalized; low pay, part-time, few benefits; U.S. attempting workfare: results mixed	Struggling to remain competitive; little done in terms of job training; widespread anxiety about effects of Free Trade agreement; recent workfare initiatives

difficulties but, with help, may be able to live autonomously or in group homes. Included are individuals who have been abused or institutionalized and require a period of time in transitional housing before they are ready for a self-reliant existence; (e) people who need continuous residential care in an institutional setting. With counseling and appropriate services, some may be able to move to half-way houses, supportive or sheltered housing.

WHO IS HOMELESS?

Older Skid Row males, some of them alcoholics or substance abusers, are homeless. So are the ubiquitous "shopping bag ladies," along with people who have been deinstitutionalized. These stereotypes, now engraved in the public consciousness by lurid press accounts, are misleading. Homelessness affects a heterogeneous assortment of people. Among these are: low-income single mothers, battered women with children who have fled their homes, workers displaced by economic change, runaway youths and abused youngsters, elderly people on low fixed incomes, those who suffer physical and mental health disabilities, substance abusers, people who are transient as a result of seasonal work, domestic strife, or personal crises, recent immigrants, refugees, and Natives (i.e. aboriginal people) who have migrated to the city in order to find work and to escape problems, ex-prisoners and those recently discharged from detention or detoxification centers or mental hospitals.

Many of these individuals have little in common except their lack of shelter. Case workers describe them in this fashion: they have multiple problems – physical, mental, social, or economic difficulties that cause them to be

1 Staff and residents of a women's shelter in Toronto, 1995
(Photograph courtesy of Joyce Brown)

marginalized. Many are substance abusers. Some are both disabled and unemployed; others have psychiatric problems that lead them to drug dependency. A considerable number have been in prison or other institutions. When they are suddenly returned to the community they have few social skills or supportive networks. Many are functionally illiterate and spend their lives moving from city to city, taking whatever temporary jobs they can find. Often they are isolated and have nowhere to turn for help.

One way of describing these disparate individuals is in terms of their characteristics; namely, gender, race, age, education, income, health status, and the length of time they have been without adequate shelter.

Gender

In the United States at least half of homeless persons are single males. In large cities the incidence of homelessness among women is higher, because of the lack of shelters for women as well as the prevalence of poor female-headed households with insecure housing arrangements. In New York City 46 per cent of the shelter population are women and female children. Of those vulnerable to homelessness a significant majority (58 per cent) are women. The U.S. Conference of Mayors found that 36 per cent of the homeless population was comprised of families with children. The vast majority of these households are headed by women (Institute of Medicine 1988: 12; Reyes and Waxman 1989: 2).

In Britain, because national legislation gives housing priority to pregnant women and to families with children, 71 per cent of those certified as homeless are women and children and another 12 per cent are pregnant women without children. In Canada about two-thirds of shelter users are males. This is the result of traditional practices: hostels in the inner city are normally reserved for single men.

Race

Minorities are disproportionately represented among the homeless population in the United States. Most homeless people in rural areas are white; but in some large cities the majority are African-Americans. In western states as well as the northeast, Latinos represent a sizable minority of homeless people – approximately 30 per cent in New York City. Across the country non-whites (four-fifths of whom are black) represent about 55 per cent of the total – 2.5 times their representation in the total population (U.S. Department of Housing and Urban Development 1989: I–2). Racial minorities in Britain are four times as likely as whites to be homeless or in very poor housing. In Canada most homeless individuals are white but in several large cities disproportionately high numbers are Natives.

Age

The median age of homeless persons is approximately 35. In general, the average has decreased over time. Homeless women tend to be younger than their male counterparts. In some jurisdictions the average age of mothers with young children in hostels is 27 and three-quarters of the shelter population is under the age of 40 (Institute of Medicine 1988: 6). In Toronto, 27 per cent of people in shelters are under age 25. Operators confirm that shelter users are younger, more racially diverse, and include more women and children than was the case before 1980 (Municipality of Metropolitan Toronto 1993: 1).

Education

Homeless individuals in both the United States and Canada have more years of schooling than their counterparts of previous decades. But they do not match the educational achievements or skill levels of the general population. About half do not complete high school and a substantial percentage are functionally illiterate, unable to deal with job applications, questionnaires, or written directions (Momeni 1989: 85, 153, 198). In Britain the majority of people accepted as homeless have relatively little education – most leave school at age 15 – no marketable skills, and very poor job prospects.

Income

In all three countries homeless people, especially those who have been on the streets for some time, have extremely low incomes. Many are completely dependent on social assistance, soup kitchens, or emergency shelters. Peter Rossi found that "there are 4 to 7 million extremely poor people, persons whose income is below two-thirds of the official poverty line and who are thus at high risk of becoming homeless." He reported that this "pool of extremely poor persons [with incomes under $4,000] that the literally homeless are drawn from has increased enormously [i.e. doubled] since 1970" (Rossi 1989: 79–81). A nationwide random sample found that Americans with annual incomes below $20,000 are almost three times as likely to be homeless during their lifetimes as those with incomes above $20,000 (Link *et al.* 1994: 1910).

Health status

About one-third of those who are homeless for more than a few months have significant psychiatric problems. At least 30 per cent are abusers of alcohol or drugs, and about two-thirds previously have been institutionalized in mental hospitals, detoxification units, prisons or jails. At shelters for adult men only one-third of the residents do *not* have illnesses or injuries. The most common health difficulties are respiratory illnesses, dental problems, blackouts, eye, ear, nose, throat problems and other illnesses associated with alcoholism (e.g. hallucinations, seizures, and delirium tremens). Garrett and Schutt's studies of Boston's Long Island Shelter found that the frequency of ill health increases with age and with length of homelessness, is greater among men than women, and is somewhat higher for veterans than for non-veterans (Schutt 1992).

Duration of homelessness

The nature and extent of difficulties encountered by homeless people are related to the length of time they have been without adequate shelter, which in turn depends on their location. In Britain those who qualify for aid generally are not without accommodation for long periods; the law requires local authorities to immediately provide temporary housing. The situation in North America is quite different. A nationwide sample (Link *et al.* 1994: 1910) found that "a majority had been homeless for more than a month." Rossi's study of Chicago discovered that roughly one-quarter are homeless for less than three months, one-half have been without shelter from four months to two years, and the remaining quarter have been on the street for more than two years (Rossi *et al.* 1987: 1336).

CONDITIONS SURROUNDING HOMELESSNESS

The social, economic, and political conditions surrounding homelessness are similar in Britain, Canada, and the United States. Responses by government and third sector representatives are different, however, as they are shaped by historical trends or precedent, by ideological positions, and by perceived economic, social, and political constraints. All three societies are grappling with such issues as globalization, deindustrialization, shortages of affordable and adequate housing, and growing disparities among regions and between deprived inner city districts and wealthy urban or suburban areas; these approaches are summarized in Table 2. In the 1980s and early 1990s, while seeking to reduce the national government's role in housing provision, they experienced substantial increases in the number of people without adequate shelter.

Broadly construed precipitating factors include economic shifts, government policies, and demographic changes. The principal economic changes, resulting from globalization, are manifested in a loss of manufacturing jobs, increased automation, and growth in involuntary part-time, low-paid employment in unskilled service sector jobs, along with streamlining of firms, accompanied by lay-offs, redundancies, and decreases in real wages.

Public allocation policies mirror a shift in budget priorities by administrations determined to cut spending on housing and social services, as they face rapidly escalating debt burdens. Government housing policies emphasize privatization and tax subsidies for middle- and upper-income home-owners. The result is an acute shortage of low-rent public housing, a depleted inventory of private rental units, and reduced shelter options for low-income groups.

Social policy changes are indicative of a shift away from a welfare state consensus and a resolve to decrease spending on assistance programs. These trends are evident in increased numbers living in poverty, declining social benefits relative to the cost of living, and an inability to afford available housing. Demographic shifts are apparent in the form of smaller households – many of them made up of seniors or single mothers with young children – and in the prevalence of non-family households. Widespread gentrification in downtown areas led to the loss of single room occupancy (SRO) units, the conversion of rentals to condominiums, and residential displacement. Changes in housing need attributable to demographic trends influenced demand for affordable shelter at the same time as supply – especially of small, low-cost units – contracted and prices rose.

SUMMARY

Because the act of defining homelessness in public policy frames the response by both government and the voluntary sector, it is vital that broad definitions are used to ensure that those most in need of assistance, including those at risk, are not excluded from public programs. At the same time, the population should not be so universally described that there are effectively no boundaries or constraints on social spending. The definition selected may help to focus public expenditures on precluding homelessness – a far more cost-effective approach than attempting to deal with people who are already on the streets. (This should not be construed as an argument for ignoring those who are absolutely homeless.)

A definition which satisfies these criteria includes those who are absolutely, periodically, or temporarily without shelter, as well as those who are at substantial risk of being on the street in the immediate future. Overcrowded, inadequate, or extremely expensive dwellings (relative to household income)

Table 2 Summary of approaches to homelessness in the three countries

	Britain	*United States*	*Canada*
Definition of homelessness	Broad; includes those at risk (some singles, psychiatrically ill, pregnant women, older people). Housing as of right widely accepted	Narrow; usually limited to absolute homelessness. Housing as of right generally not accepted	Located in between the American and British approaches. Includes some who are at risk (battered women)
Number of homeless people	7/1,000 (official figure); 25/1,000 in inner London	3–6/1,000 literally homeless	5+/1,000 (based on national shelter survey)
% of total population in poverty (using official poverty line in each country)	14.0%	14.5%	13.5%
National legislation	Housing (Homeless Persons) Act of 1977	Stewart McKinney Homeless Relief Act of 1987	None
Central government role	DoE oversees local authorities, establishes policy; DHSS provides benefits	Virtually none; limited funding to state and local agencies who may choose to deal with problems or to ignore them	Authority devolved to provinces; CMHC provides some housing aid; promotes creation of non-profit housing by cities and third sector
Local government roles	Legally obligated to certify homelessness and provide housing; substantial variation among authorities depending on degree of ideological commitment	Concentration on emergency shelters; assist people after they become homeless; some cities dealt with homelessness only when courts required it	Different in each province. Municipal non-profit housing corporations develop and renovate housing and assist private (third sector) non-profits
NGO/third sector roles	Drafted 1977 Act. Third sector involved in information and advocacy/lobbying. Development of self-help and shared housing	Operate shelters, missions and food banks. Some now developing/renovating housing; a few self-help initiatives have been successful	Development of self-help and co-op housing. Use of facilitative management model. Women's housing groups active

cont.,

Table 2 cont.	*Britain*	*United States*	*Canada*
Principal actors	DoE; local authority housing officers; advocates for homeless people; ethnic groups	City governments; NGOs; philanthropic organizations; ethnic groups	Provincial/city governments; third sector non-profit housing groups; housing cooperatives
Innovative programs	Many hostels archaic; emphasis on shared, short-life, and transitional housing. Linked housing, education, health, employment schemes	Some innovative housing, including transitional homes for women; health outreach programs; useful model of public–private joint ventures	Emphasis on co-op and non-profit housing with income mix. User design and planning of group homes

which lack basic facilities do not constitute satisfactory accommodation.[4] People who are placed in emergency shelters, hostels, or welfare hotels lack privacy, security, and rights of tenure; they must be considered homeless or vulnerable. Similarly, those low-income households with a roof over their heads, but who face serious threats of violence, abuse, or eviction, require protection. Temporary assistance should be provided before they become absolutely homeless and their requirements for social supports multiply.

Homeless people represent heterogeneous populations. Those who need help range from individuals and households who simply require temporary financial assistance to persons in need of long-term care. Most are somewhere in between: they need several forms of support, counseling, and training, but can eventually become independent and take control of their own lives.

The social, economic, and political causes of homelessness are similar in Britain, Canada, and the United States. But the definitions used, the demographic characteristics of homeless people, the responses of governments (national and local), the role of the third sector, and the scope of ameliorative initiatives vary considerably among these countries. These differences are discussed in the following chapter which deals with poverty and the politics of distribution in the three countries.

2

POVERTY AND THE POLITICS OF DISTRIBUTION IN BRITAIN, THE UNITED STATES, AND CANADA

> The most compelling explanation for the marked shift in the fortunes of the poor is that . . . [we] had changed the rules of their world . . . The first effect of the new rules was to make it profitable for the poor to behave in the short term in ways that were destructive in the long term. Their second effect was to mask these long-term losses – to subsidize irretrievable mistakes. We tried to provide more for the poor and produced more poor instead. We tried to remove the barriers to escape from poverty, and inadvertently built a trap.
>
> (Murray 1984: 9)

> Mainstream discourse about poverty, whether liberal or conservative, largely stays silent about politics, power, and equality. But poverty, after all . . . results from styles of dominance, the way power is exercised, and the politics of distribution.
>
> (Katz 1989: 7)

This chapter links the economic, political, and social issues relating to poverty and shelter in Britain, the United States, and Canada. It contrasts their views on poverty and distribution and shows the effects of recent trends in the three countries with respect to unemployment and low-cost housing. Poverty and homelessness are inextricably connected: it is not possible to make a neat demarcation separating homeless individuals from the poor. People move from poverty to homelessness and back again. The issue of homelessness, then, is related to economic problems and political decisions regarding the allocation of public funds for shelter and social programs. These decisions, in turn, are tied to social values and choices. As a manifestation of poverty and of social welfare policy, homelessness can be examined through the lens of political economy; that is, the institutional framework in which policy is formed, the economic system which gives rise to poverty and shelter problems, and the politics of distribution.

SOCIAL POLICY AND THE STATE

Social policy is defined as a course of action pursued by the state to shape society and to allocate its resources, based on "the desire to ensure every member of the community certain minimum standards and certain opportunities" (Titmuss

1974: 29). This perception of social policy contains three basic elements: value judgments about providing for the welfare of citizens; economic and non-economic objectives; and progressive redistribution of resources to meet basic needs and benefit vulnerable groups.

There are several models of social policy. One assumes that individuals' needs are naturally met through two channels, the family and the private market. Social welfare policies are employed only when these support networks break down, and then only temporarily. The second model, predicated on incentive, effort, and reward, accepts a role for social welfare institutions as adjuncts to the economy, with social needs being met on the basis of merit, work performance, and productivity. A third model recognizes social welfare as a major integrated institution for providing universalist services outside the market on the principle of need. It stresses the effects of social change and the economic system while emphasizing social equality (Titmuss 1974: 31).

Countries often incorporate elements of more than one of these models. Nevertheless, a distinction can be made on this basis between countries such as the United States, in which welfare is a last resort and must be justified through means-testing, to more comprehensive welfare states such as several of those in Western Europe. It has been presumed, in these welfare states, that the scale and complexity of modern society is such that only the public sector (in conjunction with voluntary agencies) is capable of supporting an effective society-wide policy. Proponents argue that the nature of modern industrial society is such that traditional systems of support, especially family and kinship, have been weakened by rural–urban migration, an aging population, and increased mobility (Daly and Muirhead 1993: 20–21).

Social policy, which effectively defines the state's welfare role in each of these countries, has a direct link to poverty and homelessness. Poverty in post-industrial countries is related to global economic changes, the nature of labor markets, politics, and public welfare policy. Since the late 1970s, the global economy has become increasingly integrated and corporations have automated production or have looked abroad in search of cheaper costs. Workers, as a result, have found their purchasing power and standard of living eroded. A period of growing social and economic disparity has ensued.

Social policy is given substance through social spending, usually expressed as a percentage of gross domestic product (GDP). The OECD defines "public social-protection expenditures" as health care, unemployment insurance, family transfers, disability pensions and services, and old-age pensions, but does not include education. In 1990 social programs accounted for 20.3 per cent of GDP in Britain, 18.8 per cent in Canada, and 14.8 per cent in the United States. With the exception of Spain (17.5 per cent) and Greece (20.7 per cent) all countries in Western Europe spend over 23 per cent (Förster 1994). This measure is illustrative, but not fully definitive; it is necessary to look further, to see who benefits from these expenditures. In Canada, for example, 40 per cent of all social spending went to elderly individuals. The importance of these transfers is illustrated by their impact: in Canada the poorest quintile relied on social program transfers for 63 per cent of its income.

DEFINING POVERTY

Since 1980, in Britain, the United States, and Canada, the poverty rate has grown. Explanations of poverty fall into two groups, one blaming the victim, and the other blaming the system. The first set of theories emphasizes the characteristics, attributes, and behavior of poor people. Often confusing cause and effect, it is assumed that the lifestyle of those suffering from certain social conditions is the cause of those conditions. Such explanations have been used to divert attention from serious structural problems in the economy and social system.

British government statistics do not acknowledge the existence of poverty *per se*, as the needs of low-income groups are supposed to be met by the welfare state, which is intended to lift them out of poverty. Therefore it is necessary to look at Britain in comparison to other countries, to examine the distribution of income, and to trace changes over time. The 1995 edition of *Social Trends* from Britain's Central Statistical Office reports that "since 1971 the percentage of the population below half of the average income has increased from 11 per cent to 21 per cent" (Central Statistical Office 1995: 83).

In the United States people are classified as being above or below the "poverty line" using the poverty index developed by the Social Security Administration in 1964 (and periodically revised). This index, which includes only money income, is equivalent to the level (set by the U. S. Department of Agriculture) needed to buy essential food and goods. Poverty thresholds are updated every year to reflect changes in the Consumer Price Index. According to government sources, about 14.5 per cent of the population, more than 38 million people, lived below the poverty line in 1994. The poverty level for a non-farm family of four in 1994 was $15,141 (when median income was over $32,000).

Although there is no official definition of poverty in Canada, it is generally accepted that the national standard is set by the Statistics Canada low-income cut-offs, below which people are said to live in straitened circumstances. Since 1992 the number of poor families has been calculated based on a cut-off of those spending more than 54.7 per cent of income on essentials (food, clothing, and shelter). Statistics Canada varies the cut-off levels with household size and location (rural or urban). In 1994 the low-income cut-offs for a four-person household in cities over 500,000 population was $31,256; in rural areas it was $21,600.

Making comparisons is difficult, in part because different countries use a variety of definitions. There are, however, a few studies of poverty which compare countries in terms of the same standards. One of the best known, the Luxembourg Income Study (Smeeding *et al.* 1990; Förster 1994) concluded that the actual poverty rates for 1986–87 were: Britain, 12.4 per cent; United States, 18.7 per cent; Canada, 15.4 per cent. These are *post-transfer payment poverty rates*; the rate of poverty in Britain (which, among these three countries, has the highest rate of transfer payments to poor people) was actually twice as high, 24 per cent (compared to 19 per cent in the United States and Canada), before these transfer payments were taken into account (Förster 1994).

Recent studies are available at the city level as well. An analysis of London and New York found that "despite the large gap between rich and poor, the degree of social polarization associated with inequality in London is lower than in New York" because levels of welfare provision are higher and more secure for many poor people. "Except for school leavers, program eligibility is not restricted by age and family status, as is the case in the United States" (Logan *et al.* 1992: 141). Poverty in New York and, by extension, in many American cities "is firmly linked with racial, social and spatial factors, being overwhelmingly concentrated among black and Hispanic people, among single parents, and in large areas of the city" where poverty is the norm (Logan *et al.* 1992: 149).

A DEMOGRAPHIC PROFILE OF POVERTY

When comparing poverty rates for different countries, it is important to ensure that all people are counted. Yet, virtually all industrialized countries exclude institutionalized and homeless populations. Britain counts only those people on the electoral register. Statistics Canada does not identify disabled people in its income data, nor does it include Aboriginal people on reserves. It was not until the 1990s that the U.S. Census Bureau and Statistics Canada attempted to count homeless persons (Sarlo 1992: 154). It is also essential to differentiate among population groups. Not only are aggregate figures misleading, but they gloss over inequalities. When public resources are scarce it is necessary to carefully direct transfer payments to those most in need.

One way of describing changes over time is to examine poverty rates for particular groups (e.g. single mothers) and to look at distribution of income and resources among different population groups; this allows us to focus on the status of the poorest fifth of the population (i.e. those most in need). Poverty and homelessness are linked to changes in family structure. Smaller households result in a high number of small, poorer families competing for a restricted supply of affordable housing. In the United States between 1960 and 1990 the proportion of female-headed families with children tripled, as did the number of single-person households. David Ellwood believes that, "we have reached the point at which the majority of children born in the United States today will spend some part of their childhood living in a single-parent home" (Ellwood 1988: 45–46). The proportion of mothers with children under the age of 6 who were in the labor force rose from 20 per cent in 1960 to over 50 per cent in the mid-1980s. Between 1970 and 1986 the number of single mothers with children below the poverty line more than doubled. For children in households headed by single mothers the poverty rate in 1990 (55 per cent) was 4.5 times higher than the rate for children in other families (Ellwood 1993: 3–8).

Lone-parent families

In all three countries the rate of child poverty is high in households headed by non-elderly single parents, especially those headed by women: 28.3 per cent in Britain, 48.6 per cent in Canada, and 54.8 per cent in the United States (Ross *et al.* 1994: 111; Förster 1994). (These rates are after transfer payments [e.g. welfare] have been made.) Poor women who are single parents tend to be quite young, have much less formal education, and have more young children than do non-poor lone mothers. In Britain lone-parent households comprised 25 per cent of the bottom quintile group, but only 1 per cent of the top quintile group (Central Statistical Office 1995: Table 5.16).

The proportion of single-parent families in Canada rose steadily from a low of 8 per cent in 1966 to 13 per cent in 1991. Almost 90 per cent of lone parents are women; three out of four live alone (i.e. without other adults); 16 per cent are doubled up in a relative's household; and 6 per cent are staying with people not related to them. Single mothers are younger, poorer, less educated than single fathers, and far more likely to have young children in their care. Over 40 per cent of female lone parents are not in the labor force, compared with 28 per cent for male lone parents. Two-thirds of single mothers rely on government transfer payments (social assistance and unemployment insurance) as their major source of income (Statistics Canada 1992).

Women

Labor force participation rates for women are high. Over 60 per cent of married women in Canada and 71 per cent of married women under age 65 who have children under age 18 are in the labor force. However, they are heavily concentrated in the service sector (almost 85 per cent) and are often poorly paid. Women are disproportionately represented among the very poor. They are doubly disadvantaged in that they are paid less than men in the workplace, yet bear more economic and emotional responsibility for raising children when the parents do not live together. In the United States, women are paid only about 60 per cent as much as men. Almost half of the families in poverty have a female head of household. The rise in female-headed households is related to increasing rates of divorce, separation, and single parenting. The feminization of poverty results, in part, from the failure to equalize women's opportunities in the workplace and to equalize responsibilities in the family (Golden 1992). High poverty rates among female-headed households can be traced to low incomes, high unemployment, and barriers to entry or advancement. More than four in ten families headed by women are below the poverty line, compared to one in ten headed by men.

Children

Children account for 40 per cent of poor Americans, although they represent only one-fourth of the population. One of every eight children suffers from substantial food shortages. In 1992, one out of every five American children lived in poverty; projections indicate that this ratio could rise to one in three shortly after the end of the decade. Of the 18.5 million American children in central cities, 29 per cent live in poverty. In New York City, about 40 per cent of children are poor (Ropers 1991: 45). In Canada the rate of poverty among children under age 18 rose from 15.1 per cent in 1981 to 18.2 per cent in 1991.

Elderly people

Many countries, notably the United States and Canada, have concentrated much of their spending on housing and social programs for elderly people because this approach enjoys broad political support. The success of these transfer programs demonstrates unequivocally that poverty rates can be substantially altered by social policy. The incidence of poverty among the elderly declined over the past several decades because of conscious government decisions to increase benefits for this group. Nevertheless, many elderly persons are still very poor and vulnerable; three-quarters of older people in Canada are considered low-income. Retired elderly women who are dependent on government transfer payments are over three times more likely than others in their age group to be in core housing need (Canada Mortgage and Housing Corporation 1994: 4).

The working poor

Poverty among working poor households is attributable to periodic unemployment, involuntary part-time employment, paltry benefits, and low earnings, especially in the service sector where 80 per cent of Canadians work. A large proportion of the working poor are female-headed households. In Canada between 1973 and 1986 the number of working poor increased by 19 per cent for families, and 46 per cent for unattached individuals. In the United States half of poor household heads worked in 1986, over one-quarter of them full time, year round. Ellwood (1994: 147–148) characterizes them as "the poorest of the poor . . . They play by the rules but lose the game." Many are too proud to apply for welfare. At the same time they find that their low-paying jobs are disappearing and real wages are declining. They are a huge group who are not on welfare but could be tipped over the edge by forces beyond their control (Bane and Ellwood 1994).

Ethnic minorities

Race and gender are among the most significant factors in the maldistribution of income in industrialized countries. Hispanics make up 6 per cent of the American population, but represent 12.5 per cent of the total number in poverty. One-third of all blacks are below the poverty line – three times higher than the rate for whites; they have a lower life expectancy than other Americans, are almost twice as likely to die in infancy, will receive less education, are six times as likely to be murdered, and four times as likely to be imprisoned. They also earn less. Although one out of five children lives in poverty, 43 per cent of black and 37 per cent of Hispanic children are below the poverty line. Compared to elderly whites, elderly Hispanics are two and one-half times, and elderly African-Americans over three times, more likely to be poor (Ropers 1991: 45–48).

Think globally, act locally: the impact of globalization on work

The impacts of globalization are manifested not just in international finance, capital, and labor. They are felt, in a visceral sense, at the local level by people who are "downsized" or made redundant by what are loosely termed "technological advances." While some of the high-tech jobs created by this revolution are well paid, most are menial positions, offering low wages and few benefits. People often are required to work from their homes where they have no face-to-face contact with fellow workers – this saves the employer a substantial portion of overhead costs. Many are on contract: they are not, strictly speaking, employees and therefore are not offered any security. They must work split shifts to accommodate the peak demand periods of the fast food business, for example.

- A person who worked at home telephoning to find replacement teachers when regular instructors are ill, now is employed only three hours a day feeding information to the voice-activated computer which replaced her.
- Order-takers for fast food chains work from their homes, earning low pay, working part-time and split shifts. Most are not unionized and earn money only on a piece-work basis; some are required to pay their employer to rent the computers which must be used for order taking. Other firms are now accepting orders on the Internet, thus eliminating the need for order-takers.
- Technicians at the telephone company, who once diagnosed and repaired equipment malfunctions, have been replaced by a central computer system which they helped to program. The computer operation only requires a few, low-paid clerical employees.

(M. Gooderham, *Globe and Mail*, Toronto, October 11, 1995: A1)

In Britain there are enormous differences among ethnic groups, some of which only became apparent after the 1991 Census when specific data on ethnic origin and housing and economic indicators became available for the first time. The Census reveals, for example, that certain ethnic groups are much more prone to live in social housing and to be overcrowded. Bangladeshi households, for instance, are ten times as likely to be living in overcrowded conditions, in part because they are relatively recent arrivals whereas Black Caribbeans arrived a couple of decades ago and, along with Black Africans, have the highest propensity to occupy social rented housing. Pakistanis, on the other hand, though they are similar to blacks in terms of unemployment and use of private rental housing, are much more likely to be home-owners and thus make very little use of social rented housing; as shown in Table 3, Blacks are almost four times more likely to rent social housing than are Pakistanis.

Natives

Poverty rates for status Indians on reserves in Canada appear to be 20 per cent higher than the population at large (Ross *et al.* 1994: 41). Other indicators are improving, though: housing conditions got better during the 1980s for people on reserves, as did labor force participation rates, school enrolment figures, and life expectancy. In the United States (as in Canada) Natives have high rates of suicide, alcoholism, and drug abuse. According to official statistics,

Table 3 Housing and unemployment data for ethnic groups in Britain (%)

Ethnic group	*Unemployment*	*Social housing*	*Private rentals*	*Overcrowded*
Black African	17	52	20	15.1
Black Caribbean	14	45	7	4.7
Pakistani	14	13	11	29.7
Bangladeshi	16	43	12	47.1
White	7	24	9	1.8

Source: Office of P. pulation Censuses and Surveys (1993: 284, Table 9; 1166, Table 49)

about 28 per cent of Natives in the United States live below the poverty line (twice the rate for the total population).

Disabled people

Roughly 13 per cent of the population in all three countries is disabled; well over half have very low incomes. Canadian data show that disability results in a higher poverty rate than almost any other population, including Natives (Ross *et al.* 1994: 41–42).

The remainder of this chapter deals with each of the three countries in terms of poverty, unemployment, and housing trends which are linked to the rise in homelessness during the 1980s and 1990s.

POVERTY AND UNEMPLOYMENT IN BRITAIN

> Two nations; between whom there is no intercourse and no sympathy; who are as ignorant of each other's habits, thoughts, and feelings, as if they were dwellers in different zones, or inhabitants of different planets.
>
> (Disraeli 1845/1980: 96)

Britain today is still a divided society. The income divide between the top and bottom fifth of society has widened since the 1970s. Britain's poverty gap – the amount required to lift a person above the poverty line – grew during the 1980s and early 1990s. The poorest quintile's share fell, relative to the richest quintile, in terms of original income, disposable income (after distribution of taxes and benefits), and final income. The principal reason for the decline in original income for the poor is that roughly 3 million are unemployed, and many of those working are in low-paid service jobs with poor benefits and tenuous job security. During the 1980s real disposable income declined for couples with children and for single people with children. Moreover, the wealthiest quintile increased their share of the national economic pie during the Thatcher years: during the 1980s the average annual disposable income for this group increased by 40 per cent, but there was no discernible trickle-down effect (Townsend 1993: 14, 232).

During the 1980s, a third of all manufacturing jobs in Britain disappeared. In an effort to reduce inefficiency, many businesses announced redundancies. The last hired – women, youth, and minorities – were the first dismissed. Some of the young people on the dole have never found, and may never find, a job. Over 40 per cent of those receiving unemployment benefits in Britain have been out of work for longer than one year. The highest unemployment rates (and the greatest concentration of long-term unemployment) are in the inner cities where unskilled, low-paid, manual workers are concentrated. Some council housing estates in places like Liverpool and Newcastle have jobless rates of 60 per cent or higher.

The Archbishop of Canterbury's Commission found conditions within these cities which they thought existed only in the Third World:

> Social disintegration has reached a point in some areas that shop windows are boarded up, cars cannot be left on the streets, residents are afraid either to go out themselves or to ask others in, and there is a pervading sense of powerlessness and despair.
>
> (Archbishop of Canterbury 1985: xiv)

The Inner Cities Research Programme of the Economic and Social Research Council (Buck *et al.* 1987: 122) reported "an exceptionally high incidence of unemployment among the young, the less-skilled, and the black population." In the early 1990s, 8 million people in Britain were dependent on Income Support payments, and 1.5 million had been drawing benefits for over five years. Pay differentials worsened and 3.5 million full-time adult workers earned less than the amount judged to be "low pay." During the 1980s unemployment rates doubled in the North of England, Wales, Scotland and Northern Ireland.

Because employment prospects are bleak, young people often do not feel an incentive to remain in school beyond age 15. In Britain only 29 per cent of young people obtain an upper secondary school qualification, compared to 48 per cent in France, 68 per cent in Germany, and 80 per cent in Japan. Slightly over half of British youth (age 16–19) are in full-time education or part-time education and training; this compares to more than three-quarters in Germany and France, and well over 90 per cent in Japan.

HOUSING TRENDS IN BRITAIN

Throughout the 1970s and 1980s Britain had the lowest spending level on housing in Western Europe. Total public expenditures on housing declined from £14.8 billion to £7.5 billion, as capital spending was reduced from £8.3 billion to £1.5 billion (Department of Environment 1991). The increase in mortgage interest tax relief to home-owners – which doubled during the 1980s – was equivalent to the total cutback in public spending on housing. In the early 1990s households in the top income decile received almost three times as much subsidy (in the form of tax relief) as those in the bottom decile (Maclennan *et al.* 1991: 24).

Central government passed several important measures in an effort to enforce its public policy agenda: to rein in local authorities, enhance competition, and reinforce the role of the market. Early in the 1980s the Hostels Initiative was introduced to replace large night shelters with smaller, better

2 A British Telecom tent in London's Bloomsbury district, intended to protect telephone equipment, was converted to a "mobile home." Meant to dramatize the effects of London's housing price spiral, it is advertised as a designer residence, "suitable for conversion, air-conditioned, free lighting, telephone installed, within spitting distance of all amenities. £45,000 or next offer. Apply British Telecom" (*Roof*, May/June, 1988; photograph by Tim Mars, courtesy of Shelter)

quality accommodations. Though standards improved somewhat, the number of bed spaces available to homeless single people – the principal users of hostels – declined (Anderson 1993).

The Social Security Act 1986 replaced Supplementary Benefit with Income Supports. At the heart of this alteration were four proposals: to effect a long-term cut in social security budgets; to reduce young people's expectations of state benefits; "to reinforce personal independence . . . reduce dependence on income-related benefits"; and to move toward privatization of welfare provision (Department of Health and Social Security 1985: 18–19). Since 1988 single persons have not been eligible for board and lodging allowances – used to pay rent in advance – which were replaced with housing benefits that are paid in arrears, making it difficult to put down rental deposits. The minimum age for social security was raised from 16 to 18.

The Housing Act 1988 reduced tenants' security of tenure in order to induce landlords back into the private rental market. In the following year council housing rents were allowed to rise toward market levels. During the 1980s rental subsidies to council tenants declined (in real terms) by 80 per cent. Means-tested grants were introduced to defray the cost of repairs and improvements for very poor households. Public funds were set aside for the purchase of empty dwellings and to provide hostels and "move-on housing" for single homeless people and for young people leaving institutions or foster care (Niner and Maclennan 1990: 11).

In 1990 government, housing associations, and voluntary organizations collaborated on the Single Homelessness ("Rough Sleepers") Initiative which was funded at a level of £15 million, later raised to £96 million. Designed to find emergency accommodations for 1,000 homeless people on the streets of London, this scheme produced only 230 bed spaces in its first ten months (Anderson 1993: 22). Later, the government reported that, between 1990 and 1995, about 950 places in short-term hostels and 2,200 permanent and 700 leased spaces were provided as "move-on" accommodation.

Government extended its control over municipalities while reducing its financial support. Intent on restructuring the housing supply system, central government created local authority spending limits and sanctions for overspending ("rate-capping") which limited local ability to levy taxes. Metropolitan Councils, including the Greater London Council, were abolished. Dramatic changes in tenure took place: 1.5 million council homes were sold at a discount to sitting tenants between 1979 and 1993. Local authority building came to a virtual halt; annual council housing starts in England fell from 110,000 in 1975, to 2,700 in 1991 (Department of Environment 1992: Table 6.1). Stemming in part from government cutbacks, this trend reflected a decline in slum clearance (which meant there were fewer people to rehouse), the diminished reputation of council housing's "problem estates," and decreased demand from employed tenants who met the income and other requirements of local authorities (Holmans 1991: 208).

In the 1980s the private rental market declined to the lowest level in history (8 per cent of total stock), and the role of housing associations in providing dwellings for rental and sale was markedly enlarged. In the 1992 autumn statement the Treasury announced a policy shift, indicating that £577 million would be provided – from funds earmarked for other housing programs in future years – for housing associations to buy empty properties.[1] Officials hoped that housing associations would acquire the huge stock of repossessed properties. This proved impractical. Instead they purchased units from distressed developers at discount prices.

From 1984 to 1990, while total stock grew by 4 per cent, there was a 13 per cent increase in owner-occupied dwellings, a matching decline in public and private rentals, and a 27 per cent rise in housing association rentals. The proportion of owner-occupied dwellings rose from 43 per cent in 1961 to 67 per cent in 1995. The rate of home-ownership is now higher in Britain than in the United States or Canada (which are about 64 per cent and 63 per cent, respectively). The social housing sector in Britain, however, still accounts for 27 per cent of total stock, compared with 6 per cent in Canada and about 1.5 per cent in the United States (Central Statistical Office 1992b: 146).

The combined effect of rising interest and unemployment rates created grave difficulties for low- and moderate-income home-owners. During the early and mid-1980s the national housing market enjoyed a prolonged boom, followed by a recession at the end of the decade. Building costs matched the rise in inflation – slightly over 50 per cent – while land prices increased 325 per cent. This encouraged hoarding and speculation, helping to fuel the boom and maintain artificially high prices. At the end of the decade nominal mortgage rates almost doubled, to 15.5 per cent, and the average owner's mortgage – most have variable rates – rose by 14 per cent. In 1991, one in twelve mortgage holders – 800,000 households – was in arrears, representing a 50 per cent increase from the previous year. Repossessions doubled, to 47,940, more than twice the number of new homes built by local authorities and housing associations. Some low-income home-owners lost their houses and became homeless (Department of Environment 1992: 3).

Britain has considerable problems with deteriorating stock, particularly in the private sector. The worst are the 290,000 "houses in multiple occupation" which shelter 2.6 million people: included are shared accommodations, bed and breakfast (B & B) establishments, and lodging or guest houses. Most of those living in substandard units are poor and unemployed; many are single pensioners (Niner and Maclennan 1990: 31–33).

In contrast to the inter-war years, public sector housing has increasingly accommodated low-income people, female-headed households, unemployed or unskilled workers, and ethnic minorities. In 1962, 11 per cent of local authority tenant households had no earnings (compared to 16 per cent of all households); in 1988, 60 per cent had no earnings (compared to 34 per cent of all households). Two-thirds of single-parent households live in council housing (Forrest and Murie 1990). Inner city council housing estates are plagued with problems and their occupants are often referred to as "residualized." In 1995 Prime Minister John Major decried these squalid ghettos: "There they stand: grey, sullen, concrete wastelands, set apart from the rest of the community, robbing people of ambition and self-respect . . . " (*Guardian Weekly* May 7, 1995: 8).

PRIVATE AFFLUENCE AND PUBLIC SQUALOR IN THE UNITED STATES

Because of the size and diversity of the United States the situation is not as clear-cut as in Britain. Among the principal characteristics which distinguish the American experience are a high degree of mobility, an exodus of housing and jobs to the suburbs, the lack of direct public sector involvement in shelter

3 As a result of severe shortages and social changes during World War II, the housing problem moved toward the center of the political stage in Britain. Politicians realized the value of shelter as a social safety valve and promised new homes as a reward for sacrifices made during the conflict. Inter-war housing (foreground), post-war tower blocks (background) at Rockmount Road Estate, Woolwich, London (Greater London Council 1965)

provision, and a reluctance to tolerate government intervention in social welfare issues.

Political events, such as the emergence of a strong right-wing majority in Congress in 1995, reflect population changes over the past few decades: about one-half of Americans now live in the suburbs, compared to one-third in 1960. Many, in fact, have withdrawn to gated communities with their own private security guards and surveillance systems. Private guards now outnumber publicly employed police officers in the United States. The 1990 Census found that suburbanites represent a majority in fourteen states, including six of the ten most populous (California, Pennsylvania, Ohio, Michigan, Florida, and New Jersey). Of the twenty-five fastest-growing cities identified in the Census, nineteen are actually suburbs. Of the 435 congressional districts, 40 per cent are mostly suburban (U.S. Bureau of Census 1990b). Voters in these areas are typically middle-class white property owners. Though well educated, independent, and moderate on social and civil rights issues, they have little interest in such predominantly urban problems as poverty and homelessness. Fiscally conservative, they believe that social welfare spending (except for "universal" programs like Medicare and Social Security) should be severely constrained.

Among the industrialized nations the United States has one of the highest per capita poverty rates, the most children living in poverty, the greatest gap between rich and poor, the largest infant mortality rate, one of the most severe problems of adult illiteracy – about one in five persons – and the world's highest per capita prison population. The United States ranks second to last among major industrialized nations in the rate of taxation on income. It stands last in terms of economic equality. Recent tax changes have penalized wage-earners: after the Reagan era three-quarters of Americans owed more taxes than they would have if the 1977 tax laws had remained in force. Borrowing to fund the federal deficit, accounting for about 5 per cent of federal spending in the 1960s, increased to almost 30 per cent in the mid-1990s.

Only one out of eight federal benefit dollars reaches Americans in poverty. More than 40 per cent of the poor receive no welfare, food stamps, Medicaid, school lunches, or public housing. In 1995 39 million Americans lacked health insurance and another 40 million had such inadequate coverage that serious illness could precipitate financial catastrophe. For many, incapacity in old age results in indigence. Unlike most developed countries, the United States does not provide paid maternity leave for working mothers nor does it give medical care and financial assistance to pregnant women.

Federal policies based on theories of trickle-down economics resulted in the anomaly of an expanding economy accompanied by substantial increases in poverty. During the 1980s the richest decile experienced an increase in average after-tax family income of over 27 per cent, while the real disposable incomes of the poorest decile dropped by more than 10 per cent. In 1994 the bottom quintile had a 3.6 per cent share of total income (down from 3.8 per cent in 1989) while the top quintile held 21.2 per cent of total income (up from 18.9 per cent in 1989).

Poverty in the United States became more persistent after the mid-1970s. Through the 1980s the chances of poor people escaping poverty declined (Adams *et al.* 1988). The concentration of poor persons in the nation's fifty largest cities grew by half, from an average of 16 per cent to 24 per cent between 1970 and 1980 (Bane and Jargowsky 1988). In these cities low-income blacks are five times more likely than whites to live in an extremely poor neighborhood (where at least 40 per cent of residents are below the poverty line). Table 4 describes trends in poverty and unemployment in the nation's largest cities since 1970; it shows a significant concentration of minorities in urban areas, increases in unemployment and poverty, and growing numbers of female-headed families.

HOUSING TRENDS IN THE UNITED STATES

A dilemma for poor people, even those employed full time, is finding housing which is affordable, available, accessible, and appropriate (in terms of size, location, and characteristics). Almost two-thirds of poverty-level households are renters. During the 1970s and 1980s the supply of low-rent units in the United States declined (in large part a result of gentrification) while the number of low-income renters rose dramatically; there is now a supply gap of 4.1 million low-rent dwellings. Poor home-owners, half of them elderly, also confront housing problems. Foreclosures and evictions of home-owners in the

Table 4 Poverty, income, and employment in 94 large U.S. cities (%)

	1970	*1980*	*1990*
Population as a % of U.S. population	22.5	20.9	20.1
% minority population	24.1	37.1	40.1
Unemployment rate	4.7	7.3	8.1
% employed in manufacturing	22.1	17.4	14.0
Family income as % of U.S. median family income	100.4	92.6	87.5
Family poverty rate	11.0	13.6	15.1
Population in neighborhoods with 40%+ poverty	5.1	8.1	10.8
Female-headed families as % of all families	10.4	13.8	14.5

Source: U.S. Department of HUD, *HUD User* (August 1995: 5)

1980s reached the highest level since the Depression (Leonard and Lazere 1992: 16–19; Dolbeare and Kaufman 1995: 1–2).

In the 1970s and 1980s median rents increased at twice the rate of median incomes. As a result, people were forced to double up. During the 1980s, only one-quarter of the nation's low-income renters benefited from federal housing programs. Production of low-rent or subsidized housing virtually ceased after 1982. Government rental assistance declined relative to need. In 1974, 2.2 million renter households with incomes under $5,000 received no rental assistance. By 1987, this group of eligible but non-subsidized households had grown to 3.2 million. As the poor faced higher rents and lower incomes, they also confronted reduced federal benefits, tighter income eligibility limits, and more stringent offsets for earnings.

Homelessness is influenced by a complex array of supply and demand factors affecting the nature, location, and affordability of rental housing. Urban residential markets have been altered by two opposing trends: abandonment and the return of prosperous professionals who displace existing residents. In the 1970s the population of most of the largest metropolitan areas declined – these included New York City, Chicago, Philadelphia, Detroit, Boston, and St. Louis. After 1980, with the exception of Detroit, all of these cities experienced population increases attributable to gentrification.

The loss of housing resulting from gentrification, including the demolition or conversion of about 5 million rooming house and single room occupancy (SRO) units, often took place with the concurrence of local government. New York City lost 109,000 SRO units from 1971 to 1987. Half of the stock in Los Angeles and Seattle disappeared between 1970 and 1985. Almost one-fifth of San Francisco's inventory vanished between 1975 and 1979. Denver had only seventeen SRO hotels left in 1981, down from forty-five in 1976. Between 1973 and 1984 about 18,000 of Chicago's SRO units were converted or destroyed. Virtually all of Boston's stock was lost to conversion, demolition, fire, or urban renewal (City of Boston 1986; Marcuse 1987: 426). The demise of this resource has profound repercussions because the low rents

A profile of poverty in Boston

Gross poverty rates are misleading. While the incidence of poverty in Boston is not unusually high, the rates for particular groups are substantially above the city's average. If Hispanic poverty statistics are combined with those of single parents, for example, they reveal that eight out of ten Hispanic single parents are poor and nearly three-quarters of Hispanic children are growing up in poverty. Women are three times more likely than men to be poor in the City of Boston. Only 6 per cent of singles are poor, but 41 per cent of single mothers live below the poverty line.

It is commonly accepted that virtually all poor people are on welfare. But in Boston 37 per cent have never received general assistance and one in three has not been on welfare at any time during the past five years. Fewer than one in three has received welfare continually for the past five years. Forty-four per cent of the poor people in Boston do not receive food stamps or AFDC or General Relief. Half are offered neither rent subsidies or public housing. Those who do not receive assistance pay about 60 per cent of income for rent.

Of those on welfare one-fifth were working and two-thirds would like to work if they could find jobs. Those not working tend to be ill, have sick relatives, small children, or poor mastery of English; three in ten have health problems; but at least one in four lacks health insurance of any kind. Poor people are four times as likely as the non-poor to have health difficulties and disabilities; one in five poor men is permanently disabled.

Another myth is that poor people are highly dependent on public institutions and community organizations. In fact, three-quarters were not aware of any such organizations; most do not know what groups are available to help children in trouble; and 90 per cent of Hispanics in public housing are not familiar with any community group or agency.

(The Boston Foundation 1989)

charged for SROs are affordable to people who are disabled, retired, receiving welfare assistance, or employed in low-paid part-time jobs.

During the 1970s and 1980s, only one moderately priced housing unit was produced for every three units lost through demolition, conversion, or rental increases. Less than 7.5 per cent of private multi-family stock is within the reach of poorer rental households; as a result, they are forced into over-crowded and substandard housing. Between 1970 and 1989 the number of renter households in the country increased to almost 33 million while average household size declined from 2.73 to 2.0 persons (U.S. Bureau of the Census 1989: 160). These trends are attributable to demographic shifts: more independent elderly households, higher divorce rates, increasing numbers of women in the work force, delayed marriages, and more lone parents.

With more renters in the market and fewer units available, pressure on rents grew. By 1985, four-fifths of poor renter households paid more than 35 per cent of their income for rent; for more than half of these households rental payments accounted for over 60 per cent of their income (Apgar 1989: Appendix, Table 6).

Profound changes occurred in employment and earnings during the 1970s and 1980s. Between 1979 and 1990 the proportion of full-time workers earning poverty-level wages increased by 50 per cent (from 12.1 to 18 per cent of the work force); for 18-year-olds the proportion doubled, from 22.9 per cent to 43.4 per cent. One-quarter of all full-time jobs, 24 million positions, do not pay enough to raise a family of four above the poverty line. This is a common occurrence in the high-growth areas of fast food, catering, and retailing. Wages of white male high school graduates in their prime (ages 25 to 34) fell by 21 per cent in real terms between 1970 and 1990. In the 1970s about 20 per cent of new jobs were low-paid positions (less than $7,400 in 1986 dollars); by the 1980s this proportion had risen to almost 50 per cent (Ellwood 1993: 4).

These trends exacerbated inequalities and contributed to homelessness. Even if a person works full time, year round, it is not possible, at minimum wage levels, to secure affordable housing unless there are two or more wage earners per household.[2] A woman working full time at the minimum wage would earn only 60 cents per hour more than if she remained on welfare – $1.20 per hour if Earned Income Tax Credits were received. Moreover, if she left welfare she would be likely to lose her health care benefits (Ellwood 1993: 5). Despite the growing labor force participation by women, median household income (in 1993 dollars) declined from $32,182 in 1973 to $31,241 in 1993. According to the U.S. Bureau of Labor Statistics, real average hourly earnings for American workers (in constant 1982 dollars) declined from a peak of $8.40 in 1978 to $7.41 in 1994.

The median income of home-owners is 80 per cent higher than the median for renters. While receiving tax credits and mortgage interest deductions, home-owners are not required to pay taxes on imputed rent or capital gains. This subsidy for home-owners exceeds the total spent on all direct federal housing benefits as well as local public assistance for rental payments. Those with the highest incomes and most expensive houses receive the greatest benefits. During the 1980s this subsidy more than doubled while housing assistance for low-income households was cut by about 70 per cent.[3] In 1991 households with incomes over $100,000 received an average of $5,690 in federal (cash and in-kind) benefits, slightly higher than the average amount received by households with incomes of less than $10,000. Since 1960 federal benefits have grown from 5 per cent to 12 per cent of gross national product; but, during the 1980s, the average federal benefit received by low-income households (under $10,000 annual income) declined by 7 per cent in real terms. This anomaly exists because the bulk of federal benefit programs are regressive in their application; they are allocated, not to AFDC (Aid to Families with Dependent Children) or food stamps or welfare assistance, but to Social Security, Medicare, federal pensions, and tax exemptions for company health plans. More than 60 per cent of federal benefits now go to people over the age of 64; as the population ages this outlay will grow. In fiscal year 1992 the total cost to the federal government of Social Security benefits and exemptions for corporate health plans was $90 billion.[4] By the early 1990s these benefits – representing 45 per cent of all federal spending – cost more per year than the entire federal budget in 1980 when Ronald Reagan took office (Howe and Longman 1992: 88). In 1993 the total spent on food stamps ($25 billion), welfare and family support ($16 billion), and supplemental security income ($21 billion) was exceeded by the amount of just two loophole programs: farm price supports ($16 billion) and deduction for interest ($49 billion).

Challenges of making ends meet

The most common reason that I encounter [for homelessness] is eviction for non-payment. In my area, low-wage jobs won't pay the rent for an extended duration. At $5.00 an hour ($200 per week) the take-home is around $700 per month. If the family has two children the rent will cost all of the take-home pay. The family can receive food stamps to help. But with medical, travel, utilities and all of the other basic expenses the budget is very, very tight.

If both parents go to work the take-home pay will double, but here's one glitch . . . here's a personal example . . . My wife and I both work full time at the shelter facility. We have three children at home. In winter time the children are in school so we don't have a large amount of child care expense – around $100 per month . . . In the summer we are just paying over $500 a month childcare (that's with a reduced fee) to the YMCA. If we ever lose momentum, we're in trouble. Just a simple set of circumstances could make us homeless. We are treading water all the time . . . there's little room for a breather.

(Terry McClintic, Portland, Oregon, July 1995)

POVERTY AND HOMELESSNESS IN CANADA

The Canadian safety net is more refined and extensive than its American counterpart. Universal health coverage is provided. For certain segments of the population, however, access to quality medical care and adequate housing is problematic. Natives, for example, have more health problems, poorer housing, higher morbidity rates, and shorter life spans than other Canadians (Irwin 1988). Even in major cities homeless people have difficulty obtaining proper treatment (City of Toronto 1987: 6).

Poverty is a relative concept. In Canada most poor people are not starving, nor are they forced to live on the streets. But increasing numbers, including many children and single mothers, rely on food banks and emergency shelters to meet their needs. According to Statistics Canada about 14 per cent of Canadians, 4.3 million people, including over 1 million children, are poor. Between 1979 and 1994, while median household income increased by 5 per cent, the number of poor households grew by 11 per cent, indicating that increases in income were not equitably distributed. Moreover, poverty among unattached individuals rose by 29 per cent between 1981 and 1991; at the same time, the proportion of poor elderly people dropped from 34.9 per cent to 27.6 per cent, reflecting the salutary effect of social policy being carried out through redistribution of income (Ross *et al.* 1994: 46).

During the 1970s and 1980s the total number of families below the poverty line grew by almost one-third and the number of low-income unattached individuals by half. In the early 1990s well over 2 million households (families plus unattached individuals) were in poverty. Poor female-led families represented less than one in five poor households in 1973; by the early 1990s they represented almost one in three. Their low incomes are accompanied by low educational levels and skill levels, a lack of training, very high rates of functional illiteracy, and substantial problems in finding and keeping adequate, affordable housing. While the average duration of poverty is 2.6 years, it is markedly longer for single parents (3.5 years) and for older singles, ages 45 to 60 (3.9 years) (Statistics Canada 1992: 9).

The burdens of poverty fall with particular force on children: 30 per cent of Canada's poor are under the age of 18. The Social Assistance Review in Ontario discovered that "four of every 10 persons who rely on social assistance are children" (Ontario Ministry of Community and Social Services 1988: 48). A number of direct relationships between low-income and ill health have been documented: poor children are 1.7 times more likely than other children to have a psychiatric disorder, 1.8 times more likely to die at an early age, and 2.2 times as likely to drop out of high school; "higher drop-out rates are one of the ways in which the cycle of poverty is perpetuated" (Ross *et al.* 1994: 67).

Poverty is geographically concentrated. Across Canada regional inequality increased in severity after 1984 and the proportion of low-income people was roughly twice as high (39 per cent) for inner city residents as in suburban districts. Locational disparities are an important issue in Canada. One of the commonly used indicators – though perhaps only symptomatic of more structural problems – is the level of income and poverty in various provinces. The highest rates of indigence are found in the Maritime Provinces; in Newfoundland more than one-quarter of all residents are poor. While regional comparisons are interesting it is more useful to examine the rates of poverty for various sub-populations as shown in Table 5. In recent years poverty has become more pronounced in urban areas, among young families, lone-parent households, and in households where one or more people are working.

In addition to the problems of low-income and unemployment, those living below the poverty line must contend with difficulties inherent in the health and welfare systems. The welfare system offers little incentive to engage in work or job training. In expensive cities like Toronto, Montreal, or Vancouver, most of the welfare payment must go toward housing. As prices escalated during the 1980s, rental stock was reduced, demand increased, and rents rose. An Ontario study found that 87 per cent of single welfare recipients spent more than half of their incomes on housing. An average single employable person on welfare, living in a rent-controlled bachelor or studio unit, had only

Table 5 Poverty rates in Canada, 1990 (%)

Children of single mothers	63.8
Single mothers with children under age 18	56.7
Natives	54.2
Single women over age 65	53.8
Unattached youths under age 25	50.1
Single adult women	44.5
Unattached individuals	37.5
Children of single fathers	23.1
Children under age 16 (all families)	16.9
Children of couples	10.6

Source: Statistics Canada (1992)

$47 left per month after paying for housing. As a result, many individuals "are locked into the hostel system because they are not able to pay the first and last month's rent usually required by landlords" (Ontario Ministry of Community and Social Services 1988: 58).

Ironically, while the Clinton administration urged Congress to follow the Canadian model of health care, the system in place in Canada is drifting southward, seeking to emulate American cost-efficiency and privatization. Over the period 1977–1993 federal spending on health declined from 3.2 per cent to 1.8 per cent of GDP. A central tenet of privatization is to shift power to the private sector: this principle is reflected in reduced public spending as a proportion of total spending on health, which declined from 77 per cent in 1977 to 72 per cent in 1993. By now, most health care in Canada is provided by private institutions which receive public funding. It would be misleading to suggest, however, that transfer of authority to the private and voluntary sectors enhances democratization; in many, if not most, cases, consumers of health care services within these institutions have little opportunity to participate in decisions affecting their well-being (Armstrong and Armstrong 1994).

Making do

In the mid-1990s, 1.15 million people in New York City (one in six residents) received public assistance, representing an increase of 41 per cent over five years. The mean income of the poorest tenth of the population in 1990 was $2,983 per year ($248 monthly).

A single adult on disability in New York City receives $550 per month plus $115 in food stamps. Rent and utilities (for a basic single room occupancy unit) amount to $150 monthly, leaving less than $125 per week for food and all expenses (*New York Times Magazine*, November 20, 1994: 55).

A single adult in Toronto receives $520 per month in welfare assistance, leaving a total of $170 per month (slightly over $5.00 per day) for subsistence, after paying about $350 per month rental for an SRO unit. This level of welfare assistance (representing a reduction of 21.6 per cent in late 1995 by the newly elected Conservative government) is still the second highest in Canada. In New Brunswick the total monthly welfare payment to a single adult is $267 (1995 figures).

RECENT SOCIAL AND ECONOMIC TRENDS

In Canada, jobs in extracting and processing raw materials, such as in the pulp and paper, mining and petroleum, oil and gas industries, now are surpassed by jobs in transportation and communication, where growth rates are among the world's highest. But workers in manufacturing and traditional resource industries face lay-offs and often lack the skills needed for new positions. In addition, the shift to service jobs has resulted in reduced purchasing power for lower-paid workers. The Economic Council of Canada found that "virtually all of the recent employment growth has involved either highly skilled, well-compensated, and secure jobs or unstable and relatively poorly paid jobs," most of which are in large urban centers. The shifting economy has also exacerbated disparities between cities and rural areas.

An important consequence of global economic trends, deindustrialization is manifest in declining industrial manufacturing and the decentralization of industrial activities (to developing countries or to lower-cost regions) and of back office functions (to lower-cost sites on the urban fringe). But it is not just factory jobs which are in jeopardy. Secretaries and clerical workers are being replaced by computers, receptionists by voice mail, librarians by CD-ROMS, bank employees by money machines, and many middle managers are being declared redundant because they no longer have employees to supervise.

Increasingly, the working poor in Canada are at risk of losing their shelter. This trend reflects global economic shifts. In recent years, the growth in part-time (including temporary or short-term) jobs was four times the increase in full-time employment. The number of workers on involuntary part-time status tripled during the 1980s. Statistics Canada found that from 1981 to 1984 more than half the people hired to replace dismissed employees took pay cuts. During the 1980s the purchasing power of a minimum-wage worker fell by one-third. In 1991 the minimum-wage represented less than 70 per cent, compared to 84 per cent in 1980, of the urban poverty level for a single person (Ross 1990: 12). Workers filling these part-time positions are younger and they earn lower pay and receive fewer benefits than full-time employees performing similar jobs. Opportunities for training and promotion are minimal and job security is limited or non-existent. More than half of the low-income households in Canada are working poor; 56 per cent are headed by a member of the labor force and 27 per cent by a year-round worker. Households led by part-time workers are five times more likely to be poor than those headed by full-time workers (Ternowetsky and Thorn 1990: 36–39; Economic Council of Canada 1991: 6).

Critics of the social security system in Canada believe that it has not kept pace with these social and economic trends. Family allowances were partially de-indexed in 1984 and have not matched increases in inflation. The federal government recently introduced claw-backs in unemployment insurance and the Canada Pension Plan. While real incomes and purchasing power rose during the 1960s and 1970s, both indices dropped during the 1980s while average taxes rose.

Social and economic indicators are closely linked. Significant changes in job patterns affect incomes, unemployment rates, the number receiving welfare benefits, and the incidence of homelessness. Part-time employment increased from less than 11 per cent in 1975 to 17 per cent of all employed persons in 1994 (Statistics Canada 1990: 16–19; 1995: A-4). Labor force participation rates dropped to 65.3 per cent in 1994, reflecting reduced job prospects, particularly for people without higher education or specialized skills. Structural economic shifts displaced workers and at least one-quarter of them have long-term problems finding new jobs: the average duration of unemployment grew from less than sixteen weeks in 1981 to twenty-one weeks in 1991 (Statistics Canada 1992: 3.1, 3.11; 1995: B-4).

In addition to a dramatic increase in the numbers of small households, there has been a shift of political power to the suburbs, echoing the American pattern. In the 1988 national election, for instance, the Conservatives won only nine of twenty-three seats in Metropolitan Toronto; but they captured fourteen of sixteen seats in the suburban fringe. The suburbs are emerging as a distinct political entity where voters are concerned with taxes and government spending. They are more affluent than their urban counterparts, younger and better educated; most households have two wage-earners. Whereas half of the households in the city rent, at least three out of four in the suburbs are home-owners. Suburbanites are half as likely as urban families to rely on government for wages or transfer payments. Once people move to the urban fringe, invest in a house and settle down, most lose interest in urban problems and are more concerned with economic and personal security.

HOUSING TRENDS IN CANADA

The accepted measure of housing need in Canada is "core housing need"; it refers to households unable to obtain acceptable housing with heat, running water, and customary amenities at a price not more than 30 per cent of gross household income. In 1988 Canada Mortgage and Housing Corporation (CMHC), the national housing agency, calculated that one in seven households was in core need; 70 per cent of these are renters. Almost half of single parents (almost 90 per cent of whom are women) are in core housing need, as are 41 per cent of elderly households, and almost one-third of non-elderly singles (Engeland 1990–91).

Those in core housing need are supposed to represent a priority for the federal, provincial and municipal governments. In recent years, however, the resources allocated to housing by the public sector have not been sufficient to meet the needs of the poorest quintile of Canadians. About 1.3 per cent of the federal budget was allocated for housing programs in 1990; over 90 per cent of these funds are earmarked for continuing subsidies to social housing and to Native, non-profit, and cooperative projects. Government housing production has declined since the 1970s and federal resources are intended to augment home-ownership opportunities.

Rental housing does not have priority; only about 20 per cent of total annual housing starts are rentals. As a result, problems are evident in rent levels, affordability, and housing conditions. During the past decade rents increased more rapidly than the consumer price index and almost one-third of households pay more than 30 per cent of income for rent. One in five renters live in dwellings which CMHC considers inadequate or unsuitable (Federation of Canadian Municipalities 1991).

Changes in the characteristics of tenants over time (e.g. a higher proportion of elderly people – mostly women – in public housing, more residents dependent on social assistance, and an increasing proportion of immigrants as well as single mothers with young children) reflect the residualization of low-rent social housing, changes in immigration policy, and economic shifts to relatively low-paid jobs. Also important are such trends as higher divorce rates, growing numbers of lone parents, and the changing roles of women in society and in the work force.

These shifts have been accompanied by CMHC's conscious effort, starting in 1984, to cease production of public housing and to devolve significant responsibility for housing and related issues to the provincial and territorial governments. CMHC still funds housing programs and continues to influence social housing production. The 1986 Federal/Provincial Global Agreements on Social Housing, which set out cost-sharing arrangements for non-profit, rent supplement, and residential rehabilitation programs, offer participation incentives to provinces and non-profit organizations. CMHC and the provinces share the difference between operating costs (mortgage, utilities, maintenance, and taxes) and the amount paid by tenants (based on a rent-geared-to-income formula).

SUMMARY

Homelessness is much more than a housing issue. To understand the nature and complexity of this phenomenon it is necessary to examine the political economy of the state and the nature of decisions made regarding resource distribution. Close connections are apparent in Britain, the United States, and Canada among global economic changes, poverty, unemployment, welfare policy, housing, and homelessness. In many cases those at risk are not well served by government policy decisions which tend to benefit those who are relatively well-off, vocal, and can exercise political clout.

In all three countries the rate of poverty fell during the 1960s; but this decline was reversed during the late 1970s and the 1980s as governments reduced social spending. Shifting public policy, which diminished the staying power of the poorest households, was partially responsible for a rise in homelessness, as were changes in the nature and location of jobs and in the distribution of income. In the United States and Canada, for example, the highest income group received at least nine times as much of the "national income pie" as their counterparts in the lowest income quintile (this compares with a factor of seven in Britain). The average after-tax incomes of families in all income groups except for the top quintile fell in the late 1980s and early 1990s.

From 1980 the number of poor households grew in all three countries. The rate of poverty increased for specific groups: lone-parent mothers and their children, young workers, disabled persons, ethnic minorities, and inner city residents. These indicators point to substantial differences in social policies, reflected in government transfer payments to those in the lowest income groups. The pre-transfer poverty rates are higher in Britain than in the United States; but the post-transfer rate in Britain is only two-thirds that of the United States. In Canada, the gap between rich and poor has been reduced as a result of transfer payments and income taxes. The income share of poor families almost tripled, while the share of those at the top of the income pyramid declined by about 5 per cent, as a consequence of such programs.

In all three countries during the 1980s and early 1990s, as a result of such trends as globalization and technological change, economic and social divisions became more pronounced, real wages for the working poor shrank, and the gap between housing rents and income widened. Not coincidentally, these trends were accompanied by growing numbers of homeless people.

3

THE EVOLUTION OF HOMELESSNESS IN BRITAIN, THE UNITED STATES, AND CANADA

The palace is not safe when the cottage is not happy.

(Disraeli 1848)

This chapter traces the welfare and housing roles of public and private agencies, and the ways in which programs dealing with poverty and homelessness evolved in Britain, the United States, and Canada. A number of themes are emphasized in this selective review of historical events. During the nineteenth century support for notions of social order and self-reliance was reinforced by a network of charitable organizations, settlement houses, and philanthropic builders of model homes. Those requiring relief were denied aid unless they could prove need and satisfy residency requirements. The desire for social control shaped efforts to deal with immigration around the turn of the century in North America; more recently it has affected attempts to deal with homelessness. The social attitudes and policies which evolved in Britain – a primitive Poor Law mentality, the notion of a division between deserving and undeserving poor, and the reservation of decent housing for those who could afford it – are reflected in recent approaches to homelessness in all three countries. Despite these similarities, a number of differences have become evident, particularly in recent years. These differences are explored in this historical review and in subsequent chapters.

THE IRON HAND OF THE POOR LAWS: HOMELESSNESS AND SOCIAL REFORM IN THE NINETEENTH CENTURY

In the giving of relief, the public should impose such conditions as will help the individual and the country at large. Every penny given that helps to make the position of the pauper more eligible than that of the other workmen will encourage laziness.

(Poor Law Commissioners 1834)

Homeless or destitute people in nineteenth-century Britain had few choices: they were institutionalized in workhouses which were administered in punitive fashion by the Poor Law Board or they were ejected under the Vagrancy Acts. Unpleasant conditions and regressive regulations encouraged people to find gainful employment or to emigrate. Public attitudes and the position of politicians gradually changed as a result of the investigative work of sanitarians and public health reformers who explored the causes of epidemics and the factors linking urban poverty, disease, and poor living conditions.

Starting in the 1860s charitable groups built model homes, tenements and lodging houses for those who could afford the rents. While ensuring a 5 per cent return to investors, these philanthropic capitalists used housing as an educational device, attempting to impart middle-class values to members of the working classes. The deserving and undeserving poor were clearly distinguished: the former were dealt with by such private groups as the Charity Organisation Society, the latter by the Poor Law Board. Octavia Hill founded a movement that, like the Charity Organisation Society, was widely replicated in American cities. She felt that the working class needed education and spiritual uplift, not simply better accommodations.

A housing reformer's view of her tenants

The people's houses are bad, because they are badly built and arranged; they are tenfold worse because the tenants' habits and lives are what they are. Transplant them tomorrow to healthy and commodious houses, and they would pollute and destroy them.

(Octavia Hill 1884: 10)

The prevailing view, attributing poverty to individual defects, was not challenged until all social classes were subjected to repeated outbreaks of cholera and typhoid. Exhaustive examinations of the causes of these epidemics revealed links among poverty, housing, public health, and the water supply. A gradual assertion of government powers brought about housing reforms and, eventually, better living conditions for the poor. Prompted by reformers, the state assumed a larger role in administering the Poor Laws during the final thirty-five years of the nineteenth century. While government intervention increased, a genuine effort was made to respect home rule. It was also acknowledged that there were limits to the capacity of private organizations. In 1875 Octavia Hill conceded that the combined efforts of philanthropic groups over ten years housed only 5 per cent of those needing shelter.

In nineteenth-century America, support for notions of social order and self-reliance was reinforced by a network of charitable organizations, settlement houses, and public schools. When combined with parochialism and a suspicion of foreigners this resulted in a denial of homelessness. Those who required relief were refused aid unless they could prove need and satisfy residency requirements. Industrial change and practices, immigration, economic crises, and the Civil War all contributed to homelessness after mid-century. Though industrialization came later to the United States than to England, the effects were similar. Rapidly expanding after 1865, manufacturing and construction created a huge demand for workers. Many came from rural and small-town America. They filled lodging houses in every major

4 Late nineteenth-century model dwellings, Noel Park, London (Public Housing Administration historical files, U.S. National Archives and Records Service, Washington, D.C.)

5 Model homes, New York City, following the British example; late nineteenth century (U.S. National Archives)

city. But cheap labor was still needed. A steady stream of British, Irish, and Germans filled this void, followed by Italians, Greeks, and Eastern Europeans at the end of the century.

Though many were rural peasants, immigrants frequently remained in the ports where they disembarked. Men were recruited for casual labor pools, joined work gangs on railways or canals, or found jobs in factories, while women and children labored in sweatshops or at home in tenements. Immigrants generally had no savings, no urban experience to draw on, and often did not speak English; as a result, they were vulnerable to economic crises and to the wiles of speculative tenement owners.

The cities which received the migrants and immigrants were chaotic. New York's population grew from 312,000 in 1840 to 3.4 million in 1900. Chicago mushroomed from 4,000 residents in 1836 to 1.8 million in 1900, and 2.2 million in 1910. The country's population rose from 31 million in 1860 to more than 75 million in 1900. Despite the rapid growth of cities there was virtually no state or municipal government. Garbage collection, zoning, and building regulations were non-existent until the twentieth century. Sewer systems, if they existed, were primitive and ineffectual. Open drains were a common sight. Water supply was either completely lacking or was available only at unsanitary communal water taps. Rooms were small, lacking heat and light. Often they had to be shared with other families or with itinerant lodgers. Epidemics raged unchecked through these slums. Mortality rates rose dramatically around mid-century, lending credence to the claim that extreme congestion in immigrant quarters helped the rapid spread of infectious diseases.

In times of economic distress some people became "tramps," searching for employment across the country. Often working on a seasonal basis, they harvested crops or timber. Those who remained in the city were put up overnight in police stations. Jacob Riis noted that New York City's police department furnished lodging to 435,000 people annually (Riis 1902).

Poverty in Canada, as in the United States and Britain, was seen as evidence of personality defects. There was virtually no public role in social welfare. Offered in the form of food, clothing, and wood or coal – but not cash – relief was dispensed by private charities only in dire emergencies. As a last resort some municipalities offered assistance on a one-time, emergency basis. It was generally accepted by both private and public agencies that assistance, if dispensed too freely, would induce dependency (Guest 1985: 35). Homelessness was attributed to thriftlessness, incompetence, intemperance, and immorality. When there was danger of widespread unemployment, public officials and civic reformers relied on land settlement in the far North or West of Canada as a social safety valve and as a means of populating uninhabited areas.

THE MOVEMENT TOWARD COLLECTIVE SECURITY

After mid-century in all three countries reforms were stimulated by economic depression, by revealing social studies, and by the growing realization that laissez-faire policies were no guarantee against economic downturns. Also

6 Homeless man in New York City, 1890s (Jacob Riis)

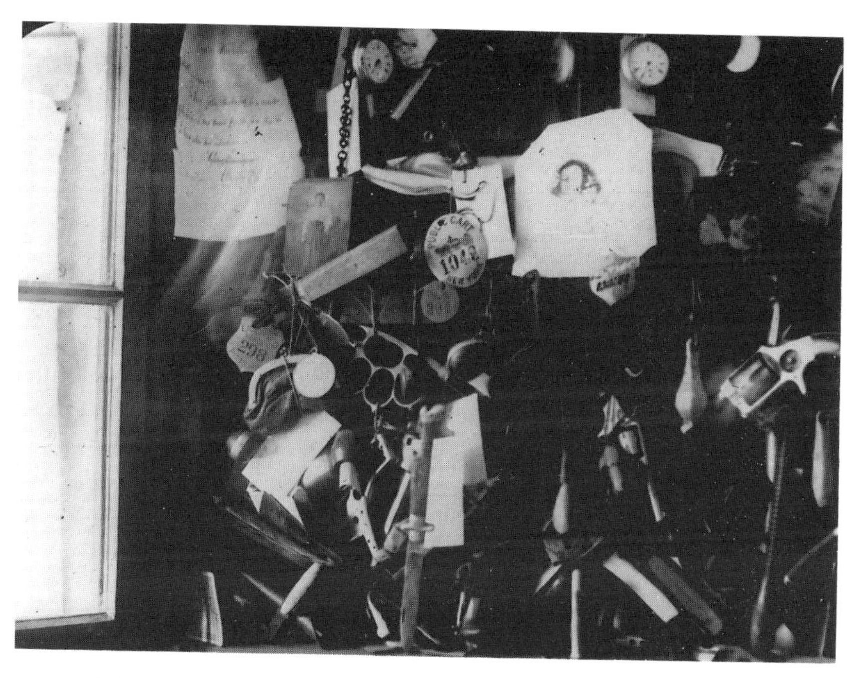

7 Lodging house residents were required to check their valuables and weapons before entering. This photograph from New York City shows brass knuckles, knives, and pistols, along with a vendor's license and a push cart permit (U.S. National Archives)

8 New York City police station lodgings in the 1890s. On an average night about 14,000 men used lodging houses in New York City; another 400 people slept on wooden floors in police stations (Jacob Riis)

important were the debates and changes initiated by new social institutions. The start of the settlement house movement in London (1884), for instance, accompanied the first steps toward emancipation of women and a greater awareness of the condition of poor people in cities.

The gap between classes was dramatic in the last quarter of the nineteenth century. By the 1880s there was a severe housing shortage in British cities, land prices rose rapidly, and thousands of slum dwellers were displaced by railroad and other construction projects. Land for housing was scarce. Because they had to remain within walking distance of their jobs, the urban poor were forced to live in substandard, cramped, but expensive quarters.

Parliament responded with the Royal Commission on Housing for the Working Classes (1884–5) which concluded that the nation's major shelter problem was overcrowding, not sanitation. Stressing the necessity for government involvement in housing production, this report made a definitive case for the inability of private enterprise and philanthropic capitalism to cope with the problems spawned by the Industrial Revolution. Subsequent legislation (the Housing of the Working Classes Act, 1890) reflected the influence of middle-class reformers arguing the merits of gradualist socialism. It was becoming clear to some that poverty, low wages and unemployment, the casual and seasonal nature of jobs, and the immobility of the poor were linked with overcrowding, land and housing shortages, and high rents.

In the United States after mid-century, reactions to immigration ranged from acceptance by industrialists, to rejection by nativists, and attempts by social and political reformers to shape newcomers in an American mold. Employers initially supported open immigration from Europe in order to assure an ample labor supply. Different ethnic groups were hired to work alongside one another. Employers believed that, if workers did not speak the same language, they were unlikely to join forces against their bosses. Discrimination was widely practiced. With radicalism on the rise in late century, employers began to question the efficacy of unrestricted immigration. Working-class locals had a jaundiced view of immigrants who threatened their job security. Workers were vulnerable as a result of economic downturns and, after mid-century, they supported a number of nativist movements. Unemployment exceeded 30 per cent in 1873 and, in the Panic of 1893, it was estimated that 900,000 were jobless (Ringenbach 1973: 38).

Close bonds between voluntary associations in North America and their models in Britain were evident in the latter half of the century. Philanthropic capitalists and reformers crossed the Atlantic to monitor the work of Charles Booth, Edwin Chadwick, and Octavia Hill. Model homes in American cities and the sanitary codes and public health regulations for New York and Boston came from London. Proceeding from a belief that the deserving poor could be aided by privately initiated education and training schemes, leaders of the Charity Organisation Society argued that relief should be offered only "to those whom it [the COS] can control" (Lowell 1884). Friendly Visitors from the COS attempted to instill middle-class values in their clients among the deserving poor (Daly 1989: 400).

Starting in 1886, young university graduates founded settlement houses in Canada and the United States, modelled after London's Toynbee Hall, to assist immigrants in assimilating. Despite their intent to "make their settlement in the slums an outpost of education and culture," volunteers, many of whom found the immigrants intractable, mirrored the inherent tensions of the late nineteenth century: "In their philosophic writings, benevolence and fellowship often appeared as apologies for the fact of economic inequality; education appeared as a device for teaching manners; and reform appeared as a mechanism for retarding change" (Warner 1962: 12).

Members of the urban establishment in both Canada and the United States were concerned that their stable society was at risk from criminal violence perpetrated by "the dangerous classes." Events in the last quarter of the century seemed to bear out these fears. In each of the decades from 1870 to 1900, depressions shook the economy. Social unrest, labor radicalism, and political agitation ensued. Many people became homeless after losing their jobs and being evicted. Deplorable inner city housing conditions, including thousands of cellar dwellings, sparked reform crusades aimed at improving public health, shelter, and working conditions. In Canada, after passage of the Municipal Corporations Act of 1849, local governments began to assume social welfare responsibilities. From 1870 to 1900 the provinces had a far greater role, principally in establishing prisons and asylums, but also regulating the work of publicly subsidized private charities. This system focused on urban areas, mainly in the older cores of major cities, where very poor people and relief institutions were concentrated (Dear and Wolch 1987: 71–79).

9

10

11 Among the concerns of urban reformers in the last quarter of the nineteenth century were child labor, public health, and tenement houses. Many women and young children worked long hours in poor conditions, either in sweat shops, or as "cottage workers" in tenement rooms, making such items as clothing, cigars, artificial flowers, or mattress covers (U.S. National Archives)

---

9 Deplorable housing conditions sparked reform crusades. The New York Tenement Law of 1901 allowed inspectors to close down apartments that were considered a health risk. Occupants were evicted; no alternative accommodation was offered. By 1920 about forty other cities (including some in Canada) had followed New York's lead by introducing restrictive building codes (U.S. National Archives)

10 "Getting ready for supper in the newsboys' lodging-house." Lodging houses were used by transient workers and recent arrivals. Jacob Riis reported that nightly prices ranged from 25 cents (for a bed and a chair in a partitioned area) down to flop houses which charged 7 cents (for a strip of canvas hung between timbers) (Riis 1890/1971: 161)

THE SEARCH FOR ORDER: HOUSING AND SOCIAL REFORM IN THE EARLY TWENTIETH CENTURY

In Britain, at the turn of the century, reformers and certain politicians felt that a comprehensive approach to housing was warranted to address social ills. This, however, necessitated state intervention, public subsidies, and politicization of the housing question. Ample evidence existed of a willingness to accept a role for government in social policy matters. Unemployment insurance (for certain industries), old age pensions, health insurance, workmen's compensation, labor exchanges, minimum wages, and free education were introduced before World War I. Yet there was still widespread resistance to the assertion of central power in providing working-class housing. Legislation remained permissive rather than compulsory. Financing was not provided. Local authorities and private enterprise were left to their own devices to sort out these ticklish questions.

In the United States reform in the early twentieth century had several strands: proponents of the "social gospel" sought to civilize society; advocates for good government worked to restrict the power of ward politicians who controlled the immigrant vote; and "housers" created model homes, "neighborhood units," and garden cities, all based on British precedents.

One of the central dilemmas of American life, the struggle to tolerate diversity while maintaining a unified society and preserving social control, was evident in the tension between natives and immigrants during the Progressive Era. Fears expressed by nativists seemed well-founded when it was revealed that the country's urban population trebled between 1890 and 1910.

Reformers proselytized for the adoption of European approaches to the problems of slums and poor housing. Housing discourse was linked with architecture, the garden city, and urban planning to offer solutions for these dilemmas in the older cities. The work of this reform generation was expressed through the Garden City Association of America (founded in 1906), the emergence of the city planning movement (1909), the World War I defense housing program (1917–1918), Better Homes in America (1922), the Regional Planning Association of America (1926), and the greenbelt towns of the New Deal. All of these organizations were based substantially on British models (Daly 1989: 409–410).

Canadian housing reform paralleled the course taken in large American cities. While tenement legislation was being enforced in New York City, similar regulatory measures were initiated in Toronto. Canada's population grew from 5 million in 1901 to 7.2 million in 1911, and 8.8 million ten years later. Most of the increase was in urban areas. But economic advances brought with them social unrest and anxiety about the implications of immigration for Canadian society. Some reformers, though, recognized that industrial expansion, urban growth, housing, sanitation, and public health were linked. Their research helped to pave the way for improvements in the living conditions of working people. Workmen's compensation, based on Britain's 1897 legislation, was introduced in Quebec in 1909, and in Ontario five years later. Gaining momentum during World War I, social reform forces, in collaboration with farm and labor groups, gradually began to exercise political power.

A Canadian view of immigrants

...they owe allegiance to the Greek or Roman Catholic Churches, but their moral standards and ideals are far below those of the Christian citizens of the Dominion ... It is our duty to meet them with the open Bible, and to instill into their minds the principles and ideals of Anglo-Saxon civilization.

(*The Merthodist Missionary Outlook* 1908)

WORLD WAR I AND THE INTER-WAR PERIOD

The years just prior to World War I were a time of growing optimism in Britain. Several pressing social problems had been solved in theory if not in practice. Exemplified by the garden city, the future seemed to offer the health, happiness and bounty which had been denied to the masses during the Victorian Age. This mood was shattered on 4 August 1914. The war created substantial hardships, particularly for the poor. Faced with the prospect of thousands of unemployed, ill-housed men clamoring for work, politicians felt obliged to offer returning veterans and war workers a better life after the conflict. There was, however, no unanimity regarding the way England should be reconstructed.

It was a time of disquieting unrest across Europe: strikes in Britain were followed by mutinies in the French army, and revolution in Russia. In 1917 Lloyd George established the Ministry of Reconstruction and initiated an inquiry into the reasons for industrial disruptions. The Commission found that high prices, lack of personal freedom under wartime controls, and housing shortages were the major complaints. It was widely feared that, unless bold steps were taken, England might succumb to class war (Public Record Office 1920).

England's jarring North–South divide was apparent in unemployment figures from August 1932: Jarrow, 67 per cent, Merthyr, 62 per cent, Maryport, 57 per cent, while London's rate was less than 10 per cent, and Oxford's was 6 per cent (Orwell 1937). For those with regular jobs there was a steady increase in real wages and in living standards during the 1920s and 1930s. For the others, however, it was a grim world.

Social reform did not come easily. It was, Bentley Gilbert observed, "simply a live grenade the front benches tossed back and forth in the hope that it would explode while in the opponent's possession" (Gilbert 1970: 306). The nation moved gradually, almost reluctantly, toward a consensus on social responsibility and a guaranteed maintenance system as a matter of right. The principal reform thrust was the construction of housing to address homelessness, overcrowding, and inadequate dwellings. Housing programs during the interwar period consisted of a series of discontinuous experiments. The initial policy, to provide "Homes Fit for Heroes," presumed that the state had a responsibility to intervene because of shortages and a desire to placate dissatisfied workers and demobilized troops. A second policy assumption was that working-class accommodation standards could only improve through state intervention. The third element was slum clearance and reconstruction (Bowley 1945: 182). Steps taken during the inter-war years to deal with homelessness and to improve the dwelling stock were impressive. Over 4 million new homes were built between 1919 and 1939; 1.5 million, or 37.5

12 Garden City planning applied to World War I defense housing project in New England (U.S. National Archives)

13 In the years after World War I Britain mounted a massive building campaign to produce "homes for heroes," for the returning servicemen. During the inter-war period (1919–1939) over 4 million homes were produced by both public and private builders. In 1923 King George V and Queen Mary officially opened Becontree, a large housing estate in London's East End (Greater London Council)

per cent, were subsidized by the state. Homes were well built. A standard three-bedroom dwelling had 855–1055 square feet under roof. Housing policy started to live up to the ideals of social reform: the council house became a ubiquitous feature of life in Britain by the end of the 1930s.

Serious problems remained, however; part of the population had been left behind by the inter-war social reforms. Birth rates, infant mortality, and the incidence of disease were still substantially higher in the inner cities than in the outer wards. Many slum dwellers were no better housed in the 1930s than they had been at the end of the previous century. In 1939 one-third of the people lived in good quality new housing; one-third were in older houses which lacked certain amenities but were generally satisfactory; the lowest third, though, were compelled to live in substandard dwellings, mostly in the urban slums.

For the poor and ethnic minorities in the United States the two decades after the war failed to produce substantial economic gains. As a result of automation and economic downturns, 3.5 million workers were jobless in 1921. Many, particularly families in company housing, were suddenly cast adrift without any supportive welfare system. These years witnessed both the construction of glittering steel and glass skyscrapers in major cities and depressions in the coal regions. Miners in a dozen states became homeless. Numerous attempts at union organizing were undertaken. Strikes were commonplace. Frequently management resorted to lock-outs; returning strikers found themselves jobless or were forced to accept pay cuts. Generally supporting management, the courts were not a force for social change.

In 1928 presidential candidate Herbert Hoover praised the individualism and voluntarism that had made his country prosperous: "We in America today are nearer to the final triumph over poverty than ever before in the history of any land . . . The slogan of progress is changing from the full dinner pail to the full garage." Barely one-quarter of the population shared in this bounty, however. Before the Crash of 1929 the average weekly wage was $28.00. By 1932, for those who could find work, wages had fallen to $17.00 per week (U.S. Congress 1930).

In Canada, during the decade after the war, politicians' and reformers' concerns mirrored the experience of Britain. The Royal Commission on Industrial Relations was formed in 1919 in response to widespread social and labor unrest, including conscription riots in Quebec City and a bitter general strike in Winnipeg. A number of issues caused working people to become agitated: rising unemployment was aggravated by war profiteering and then by demobilization. Inflationary price spirals quickly outstripped wages. Many expressed concern about housing supply and the atrocious quality of dwellings for returning servicemen. Because no residential construction had occurred for several years, the war exacerbated an already serious situation (Canada 1919). The federal government responded with a minimal housing program which was intended primarily to attack unemployment in the building trades. A public clamor for better living conditions in the post-war years prompted efforts to improve the lot of poor people. During the 1920s public welfare expenditures grew by 130 per cent. Given a general desire to return to normal, however, little thought was devoted to housing and public welfare reform.

PRIVATE AND STATE RELIEF DURING THE DEPRESSION

In the United States, as people lost their jobs and savings, confidence dwindled. Breadlines and soup kitchens offered free meals. Over 13 million people, one-quarter of the work force, were unemployed, and 34 million had no income of any kind. Many moved around the country seeking work. On the outskirts of virtually every city, "Hoovervilles" appeared – tar paper shacks, lean-tos and tents erected by homeless people.

While President Hoover continued to express confidence in the American "sense of voluntary organization and community service," the homeless ranks swelled. Agricultural income declined by two-thirds. Those in urban areas were not much better off. Foreclosures reached 1,000 per day before President Roosevelt stepped in to protect home-owners. Renters, however, lacked security. New York City, where there were more than 1 million jobless, recorded over 60,000 evictions during the first three weeks of 1932.

Speculating that those on relief lacked motivation, some Americans questioned whether they should be allowed to own cars, to retain life insurance policies, to vote, to maintain citizenship, or to buy liquor. Arrests were common for loitering, and even for picking through garbage. It was presumed that homeless tramps posed a danger to the social order. Most, however, were ordinary people who had been laid off and were unable to find work.

14 Paymaster's window, 1934
(Photograph by Dorothea Lange, Library of Congress)

15 Some of those who were jobless organized demonstrations against employers and the government. This photograph shows the camp of the "Bonus Army" (near the U.S. Capitol building) being burned after the protesters (17,000 World War I veterans) were routed by the U.S. Army (U.S. National Archives)

A Pennsylvania study of 31,000 unemployed men found that the average was 36 years old, native born, physically fit, with a good work record.

For the first two years of the Depression, relief and jobs were provided by private agencies, individuals, and corporations. Some communities formed self-help groups that took men to farms where they harvested food. By mid-1932, though, charitable organizations had virtually emptied their treasuries, local governments began defaulting on their bonds, and public officials reiterated their pleas for help from Washington.

The federal government's limited response was conditioned by historical precedent, by the absence of a national reform coalition and the relative weakness of labor unions, by a decentralized political system, and by the lack of *de facto* universal suffrage. American political parties depended on regional, rather than national, support networks. Parochialism was ubiquitous. Members of Congress owed their allegiance to state and local political organizations. The federal bureaucracy was not firmly established until the mid-1930s and its scope was limited. It was generally accepted that the federal government should not intervene in local affairs. When reform did occur it was in response to a crisis, normally was initiated by coalitions with ties to legislators, and frequently was undone by a conservative backlash (Weir *et al.* 1988: 19).

In 1933, when the Federal Transient Bureau estimated the country's homeless population at 1–1.25 million, Congress offered payments to unemployed persons, to be distributed through state and local welfare agencies. The Federal Emergency Relief Administration (FERA), which housed 125,000

16 Unemployed dockers wait outside the hiring hall, New York City, 1934 (Photograph by Lewis Hines, U.S. National Archives)

people in "transient camps," put men to work on public construction projects. By mid-decade it was evident that unemployment, poverty, and homelessness were pervasive and could not be eradicated by stopgap measures.

"We must quit this business of relief," President Roosevelt declared in petitioning Congress for authorization to create 3.5 million jobs for "employables." The Works Progress Administration (WPA), which succeeded FERA, began construction on public projects worth $5 billion. In August 1935, aware of the need for long-term, comprehensive measures to replace the dole, Roosevelt signed the Social Security Act which provided unemployment insurance. Each state was free, however, to determine eligibility criteria, taxes, and benefits. Gradually the social security program was enlarged to include virtually all categories of workers. This initiative enjoyed broad middle-class and working-class support. It came to be seen as "social insurance," paid for by participants' wages, as opposed to "welfare," which to many people represented handouts.

Speaking at his second inaugural, President Roosevelt lamented that one-third of the nation was "ill-housed, ill-clad, ill-nourished." Replacing the housing arm of the Public Works Administration, he created the U.S. Housing Authority to build low-rent public housing in order to provide shelter for the poor and – at least as important – to provide jobs for a beleaguered construction industry. Despite these efforts, the country still faced severe economic problems at the end of the decade. In 1939 the U.S. Department of Agriculture initiated the nation's first food stamp program to feed the hungry

17 Migrant mother, 1934 (Photograph by Dorothea Lange, Library of Congress)

and homeless. The gross national product in 1940 was still below the level achieved prior to the Crash eleven years earlier.

The fact that the United States initiated welfare programs should not, however, be construed as a wholehearted endorsement of universality. Interest groups shape public policy. Social policy, as a result, may lead to systematic exclusion. In the history of American social welfare policy, gender, race and class are intertwined. Much of the agenda for early welfare and pension legislation was based on the accepted notion of separate spheres for women,

18 In 1934 dust storms ruined 100 million acres in the Plains. Thousands of "Okies" and "Arkies" lost their farms and began the trek to California (U.S. National Archives)

19 The federal government dispatched the Civilian Conservation Corps to assist homeless people in the wake of floods. Forrest City, Arkansas, February 1937 (Photograph by Walker Evans for the Farm Security Administration, Library of Congress)

20 A common form of housing for poor, single men during the Depression was the rooming house or the boarding house. Residents of this establishment in Birmingham, Alabama paid $30.00 per month for room and board in 1936 (Photograph by Walker Evans for the Farm Security Administration, Library of Congress)

21 Squatter camp, 1934 (Photograph by Dorothea Lange, Library of Congress)

22 Washington's infamous "alley dwellings," which were demolished in slum clearance campaigns and replaced by public housing (Farm Security Administration photograph by David Meyers, July 1939, Library of Congress)

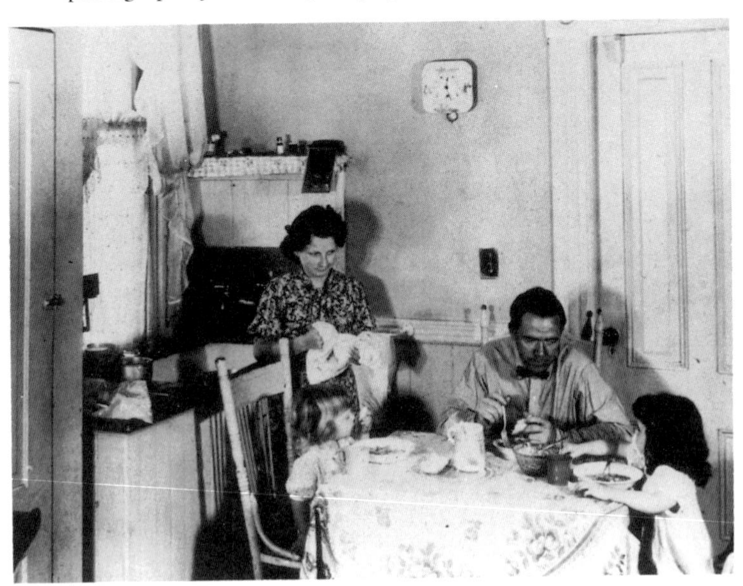

23 (Above) San Francisco and *24* (opposite, Pittsburg). Between 1935 and 1939, 88,000 public housing units were created. Intended for the deserving poor, these projects were seen as way stations on the road to independent living (U.S. National Archives)

24

whose dependency was enshrined in provisions for aid to dependent children in the Social Security Act of 1935. Gordon and others make a compelling case that gender is a major organizing principle of the welfare system and that welfare policies are based on a double standard for men and women (Gordon 1990: 10; Weir *et al.* 1988). Similarly, blacks were excluded from early legislation of the welfare state: 60 per cent of blacks (domestic workers and farm laborers) were not covered by the Social Security Act, despite the intervention of the NAACP with the drafters of the legislation (Fisher 1993).

Canada's economic decline during the Depression was as devastating as that of the United States. From 1929 to 1933 the gross national product dropped by 42 per cent. Income from agriculture in the three Prairie provinces (Manitoba, Saskatchewan, and Alberta) in 1932–1933 was only 6 per cent of the level in 1928–1929. The effects of this downturn spread throughout the economy, resulting in a loss of jobs in shipping, manufacturing, and on the railroads.

At the beginning of the Depression no public welfare system was in place. People were presumed to look after themselves by relying on savings, relatives, and private charities. Most people regarded acceptance of relief as an admission of failure. Frequently, if compelled to apply for welfare, destitute families were ostracized by their neighbors. By 1933, however, when official unemployment reached 26.6 per cent, more than one in five Canadians sought public assistance. Local and provincial governments were unable to provide much aid because tax revenues had declined precipitously; a number of municipal and provincial governments defaulted on their debts. Federal authorities urged local government to minimize relief expenditures. When Ottawa became involved it built public housing which was primarily a pump-priming device and could be afforded only by the top fifth of the population (Bacher and Hulchanski 1987: 151).

As economic conditions worsened, fears of social unrest intensified. In asserting the need to maintain order, the Bennett government moved decisively against labor and Communist militants. More than 7,000 residents were deported in 1933. All foreign-born who applied for relief in some cities were obliged to report for deportation.

In order to reduce federal expenditures the government closed the relief camps in 1936, cut relief grants by 25 per cent, and coerced homeless men to work on farms and on the railroads. These policies were echoed in the large municipalities. Civic leaders feared that urban centers would become magnets for unemployed single men during the winters. A rag-tag army of 1,800 single homeless men, organized by the Relief Camp Workers' Union, set off for

Prime Minister Bennett's view of homeless men (1935)

At least 200,000 heads of families, probably more, are still receiving unemployment relief . . . over 20,000 able-bodied employable men are in relief camps . . . hordes of homeless young graduates of the business and technical schools, and of the colleges and universities, unplaced, idle, are drifting to dependence.

(Brown 1987: 45)

Ottawa to demonstrate against federal inaction on unemployment. Their march ended abruptly in Regina on July 1, 1935 when the Royal Canadian Mounted Police forcibly halted the trek and provoked a riot. In response to these events, and mindful of the need to create jobs, the federal government passed the Dominion Housing Act in 1935 which authorized $20 million in loans and helped to finance 4,900 units over three years. In 1937 the Federal Home Improvement Plan subsidized interest rates on rehabilitation loans and the National Housing Act helped people to purchase homes and provided for construction of low-rent housing.

The Rowell-Sirois Commission's report on the Depression years

Canada's political, public finance and economic organizations were not adapted to deal with sharp and prolonged economic reverses. When a specific and coordinated programme was required, there was bewilderment; when positive action was needed, there were only temporizing and negative policies; when a realization of the far-reaching effects of the altered circumstances was demanded, there was but faith in the speedy return to the old conditions of prosperity.

(Canada 1940)

25 Seriously concerned about the risk posed by roving bands of 70,000 unemployed single men, the Canadian government established relief camps in the 1930s as safety valves for the cities. "Our purpose," wrote the camps' military commander, "is not to attempt to care for 100 per cent of the single homeless men but to reduce the numbers in the larger centres of population to the point that they do not constitute a menace to the civil authorities." This photograph shows single homeless men leaving Toronto on a train bound for work camps in Northern Ontario (City of Toronto Archives 1931)

26 Homeless and unemployed people participate in a hunger march, College Street and University Avenue, Toronto. Marchers advocated higher pay for relief workers and government action to preclude evictions (City of Toronto Archives 1934)

27 In Canada during the Depression unemployed men frequently slept outdoors when weather permitted (the bandshell outside the Parliament Building at Queen's Park, Toronto) (City of Toronto Archives 1937)

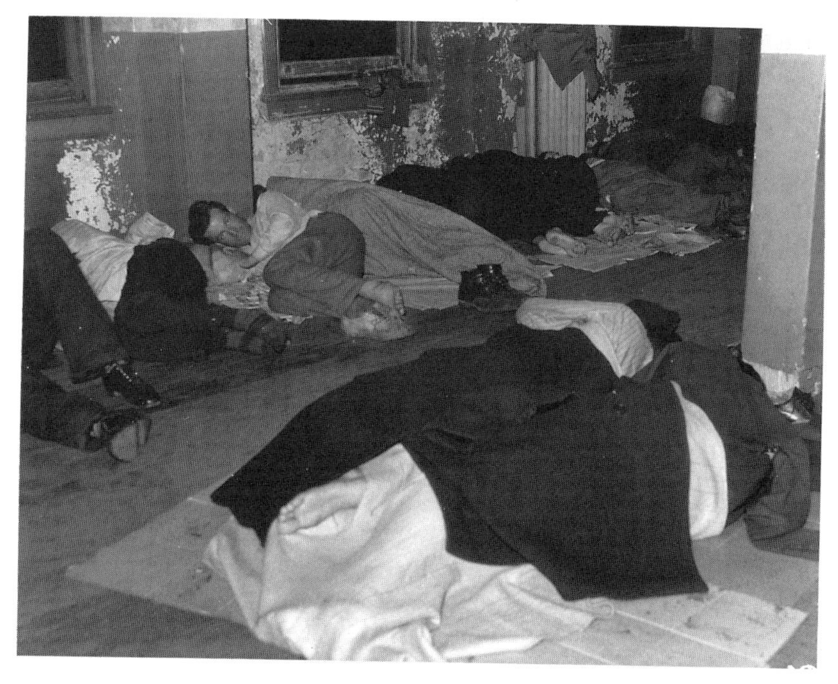

28 Flop house occupied by single, unemployed men in Toronto (City of Toronto Archives 1938)

WEAVING THE SOCIAL SAFETY NET

Federal spending in the United States grew exponentially during World War II and continued in the post-war period. The question of massive state intervention in the economy was no longer raised. More attention was given in both Britain and Canada, however, to national programs to aid the neediest during economic downturns.

The crushing economic problems of the 1930s, followed by the privations of wartime, resulted in a conviction that widespread changes were required to fill gaps in the nation's social safety net. Warning of a disturbing movement toward two Englands, the Barlow Commission on the Distribution of the Industrial Population (1940) recommended drastic reforms; these were carried out during and after the war by enlarging the state's welfare role (Beveridge 1942).

While the economist William Beveridge was constructing his model of the welfare state in Britain, Canada introduced similar social reforms, following Dr. Leonard Marsh's recommendations on health policy, employment, and training, as well as income security. Heavily influenced by Beveridge, Marsh advocated universality of benefits to be paid for by all Canadians through taxes. His recommendations resulted in the introduction of family allowances for everyone with children under 17. This was followed by old-age security in 1952 for those 70 or over, by the Canada and Quebec Pension Plans in 1965, and, a year later, by the guaranteed income supplement for those receiving

29 In the United States, after Pearl Harbor, immense amounts of resources were mobilized to build barracks for troops and housing for their families. More publicly assisted dwellings – 35,000 units – albeit many of a temporary nature, were completed in a single *month* of 1943 than had been built in any prior *year* (Naval Air Station Alameda, California, 1941; U.S. National Archives)

old-age security. The federal government subsequently brought in universal health care, and then the Canada Assistance Plan, which involved welfare cost-sharing with the provinces.

Unemployment insurance was introduced during World War II, but the government exempted seasonal industries, including mainstays of the economy: agriculture, fishing, forestry, and shipping. During the war-generated boom, relief measures were scaled back dramatically. For defense workers and others enjoying high wages during wartime, there was no need for public welfare. But regional disparities have always been marked in Canada, and certain sections of the country failed to share in the good times. There was substantial evidence of overcrowding and doubling up, especially in large cities: in Halifax, for instance, 17 per cent of households were doubled up in 1941, 26 per cent were overcrowded, and 43 per cent of the dwellings were substandard (Wade 1986: 43).

During the war, families were vulnerable to eviction. Landlords sought windfall profits from the superheated housing markets in cities and defense centers where an acute shortage of rental units caused prices to escalate. Often those without security of tenure found themselves homeless. Even when a federal government order was passed limiting repossession and eviction, tenants had only a brief respite. Pressure from the real estate lobby was so intense that the government rescinded its directive and landlords were allowed to dislodge tenants on virtually any pretext. House sales, evictions, and homelessness increased while the supply of rental units was further eroded. Some families were forced to break up, placing children in foster homes and orphanages, because housing was not available.

Early in 1944 the Canadian government appointed Emergency Shelter administrators to monitor overcrowded housing markets; migration to these areas was restricted. Municipalities used federal funding to build temporary shelters for homeless people which were intended to keep families "warm and dry . . . [with] sufficient space and essential facilities so that a good standard of health and morale [was] maintained." Officials were quick to add, however, that "these families have not been made so comfortable that they have lost their desire to search for satisfactory private accommodation" (Bacher and Hulchanski 1987: 147). The government's principal shelter activity during the 1940s was the creation of the Wartime Housing Corporation, which built 46,000 units before being converted to the Central Mortgage and Housing Corporation (later the Canada Mortgage and Housing Corporation) in 1946; CMHC's primary role was to provide home buyers with mortgages at favorable rates (Wade 1986: 43).

In Britain, although there was no specific Beveridge plan for shelter, post-war housing policy evolved as an extension of the welfare state. Millions of Britons had been made homeless by the war and millions more were living in cramped and squalid slum houses which were over 100 years old. Beveridge was convinced that the new National Health Service and the proposals for a better education system would be futile unless there was also a state-aided plan for dwellings.

The broadening of the scope of council housing and the diminution of class consciousness – as a result of collective wartime efforts and a corresponding social cohesion – were revealed in post-war legislation. A new social climate emerged in the aftermath of the war. Both parties acknowledged that public authorities would no longer build different types of estates for different social classes. Aneurin Bevan, Minister of Health and Housing (1945–1951), railed against the "castrated communities" of pre-war years, where private builders provided for one income group and local authorities for another. He wanted, instead, "to see the living tapestry of a mixed community" (Foot 1973).

THE POST-WAR ERA

Among the most significant pieces of post-war legislation in Britain was the National Assistance Act of 1948. Designed to ensure that the Welfare State reached all members of society, it obliged local authorities to provide temporary dwellings for homeless persons. Responsibility for dealing with homeless people, however, was defined in welfare legislation rather than under the housing acts. In stipulating that *temporary accommodation* was to be provided for people in *emergency* situations, the Act anticipated that:

> the demand for such temporary accommodation would come most frequently from people who found themselves without shelter by reason

> of fire, flood or similar calamity. In fact the majority of those who sought temporary accommodation were families evicted from their homes. This use of temporary accommodation for a purpose for which it was not originally intended gave rise to difficult problems.
>
> (Glastonbury 1971: 42)

Among these problems was the use of such old Poor Law institutions as sex-segregated hostels. It was common practice to shelter only women and children; men were excluded and thus families were separated. Harsh regimes were introduced by local authorities in order to encourage people to move on quickly. No provision was made for long-term housing.

In reviewing the Act, Donnison and Ungerson concluded that:

> The 1948 Act provided the wrong powers (imposing a weak obligation to provide temporary shelter for small numbers when the problem of homelessness called for strong obligations to provide permanent housing for large numbers) and these powers were in the wrong hands (resting ineffectually with the DHSS [Department of Health and Social Services] and the county social services departments instead of the DoE [Department of Environment] and the district housing authorities).
>
> (Donnison and Ungerson 1982: 271)

This resulted in a "policy gap, a classic illustration of the 'borderline' problem where policy issues straddle the areas of responsibility of different departments" (Raynsford 1986: 38). Provision for homeless families depended on the whimsy of local officials. In many cases the ball was continuously batted back and forth between welfare and housing agencies because no definitive policy initiative had been taken by local authorities. This situation was affected by a long-standing practice to reserve council housing for the fully employed, better-off members of the working class. Many housing officers and councillors looked on homeless people as idle, undeserving drifters. "I have to pay attention to the ordinary standards of decent people," observed one housing officer. "We don't want these dead-legs. They muck up the books and make life a misery for ordinary folks" (Glastonbury 1971: 106).

Meanwhile, homelessness worsened in the face of this policy vacuum. In 1969 the DHSS commissioned the Greve study, but failed to publish or to act on the results. John Greve (1971) found that a key element of homelessness was a housing shortage which caused many families to rely on temporary and insecure accommodation. He also pointed out a crucial characteristic which has now been widely accepted: the heterogeneity of the homeless population and the range of problems found within their ranks.

The lack of government response to Greve's study was illustrative of the absence of a political constituency for homeless persons. This was not surprising, given the popular belief that many were transients who failed to stay in one location long enough to vote or to find full-time work. Moreover, bona fide local residents who were already on council housing waiting lists were a far more potent political force. The two groups were in direct conflict:

> The homeless were far fewer in number, isolated from each other and generally unsupported by local networks. The process of becoming

homeless could often lead them through several short-stay arrangements, and could . . . result in them crossing local authority boundaries and being identified as "outsiders" with no claim to housing in the area. It is hardly surprising that they lost out locally.

(Raynsford 1986: 43)

A critique of local authority practice

The small minority of homeless families with multiple social problems were already felt to cause great difficulties to the smooth running of housing estates. Rather than contemplate their departments playing a wider role, most housing managers argued that such families, even those who were former council tenants, had passed out of the range of housing possibilities.

(Glastonbury 1971: 194)

In the United States, President Roosevelt spoke of an "economic bill of rights" in 1944, pledging food, shelter, jobs, and education for all Americans; but it would be a long time before these problems were attacked in comprehensive fashion. The post-war years were marked by a concern with economic recovery, growth, overseas economic expansion, suburbanization, industrial regeneration, and mass consumption.

As the Truman administration grappled with foreign affairs and the demands of economic recovery, veterans were forced to double up or to live in grossly inadequate dwellings. In an effort to dramatize their plight some pitched their army pup tents in front of city halls. Others moved into public buildings, hoping to jar the public conscience and goad the government into action. About 2.5 million families were doubled up in 1946; the U.S. Bureau of the Census (1951) found that 6 million low-income urban families were living in dilapidated and slum housing.

In 1945, Congress established the Veterans' Administration (VA) and created the G.I. Bill of Rights, providing for education, training, hospitalization, and psychiatric treatment. Congress appropriated funds for veterans' housing. While triggering explosive growth in the suburbs, these programs were most beneficial to middle-class and working-class whites. An extreme shortage of adequate, affordable shelter led to renewed demands for more public housing. The administration sought to pass housing legislation for several years after the war. Given the general tide of conservatism which swept the country, however, this was an uphill battle. The problems of cities, including urban homelessness, were neglected in favor of house and highway construction in the suburbs. Almost one-quarter of the nation's 40 million homes were defined as "substandard" in the 1950 census. In 1949–1950 the Truman administration managed to push the Housing Act of 1949 through Congress; but this legislation was vitiated by inadequate financing and local opposition.

The ambivalence of the Administration and Congress was evident in the budgets of the early 1950s. Public housing expenditures by federal, state, and

Can you afford to pay somebody else's Rent?

Read the WHOLE TRUTH about "Handout" Housing:

YOU'LL PAY $$$$
FOR *THEIR* RENT

30 (Opposite and above). In the early post-war years, when almost 3 million families were doubled up, there was substantial grassroots pressure to expand the stock of affordable rental units. But the notion of public housing was opposed as "socialist" by the real estate lobby and the American Legion. Advertisements used in Georgia asserted that "Public Housing means an end of racial segregation in Savannah!"
(Cartoons produced by the National Association of Real Estate Boards 1950)

local governments in 1954–1955 represented only 0.3 per cent of public welfare spending. President Eisenhower stressed the need for civil order and harmony, for a cooperative partnership with business, and for limited state intervention in domestic affairs.

Almost one-quarter of Americans were below the poverty line at the end of the 1950s. Only six in ten homes had basic plumbing and hot running water; half were without central heating. The top 0.5 per cent of the population held 25 per cent of the country's resources, up from 19 per cent at the end of the 1940s. Despite widespread faith in growth as an economic panacea, the country's wealth did not trickle down.

When the cities received attention in the late 1950s urban renewal was zealously pursued, particularly after the 1949 Housing Act was amended in 1954 to permit the partial use of downtown sites for commercial redevelopment. Rooming houses and single room occupancy hotels were razed to make way for new offices, luxury apartments, convention complexes, and federally subsidized freeways (National Commission on Urban Problems 1969: 86). Skid Row was shrinking in any event, reflecting a declining need for casual labor. Many of the remaining inhabitants of Skid Row were accommodated in public projects for the elderly which were built in record numbers during the 1960s and 1970s.

Public housing for non-elderly tenants was segregated and highly concentrated in urban ghettos. The total number of units produced failed to match the number lost to demolition, conversion, gentrification, or arson. Inner city abandonment continued during these decades as a result of deindustrialization. The country's suburban population increased to about 25 per cent in 1950 and 33 per cent in 1960, as upwardly mobile families sought relief from urban ills and were attracted by new developments on the city's edge. Most of those left behind in the urban core were on the margins – poor, racial minorities, old, or unemployable. Municipalities, faced with declining tax bases, encountered difficulties in maintaining physical and social infrastructure. The dire state of inner cities – in education, health, welfare, public works, and finance – received national attention after ghetto riots (National Advisory Commission on Civil Disorders 1968), the collapse of New York City's fiscal structure (1975), and the technical default of Cleveland on bond issues (1978–1979). Abetted by federal spending and tax policies, these trends encouraged a continuation of the exodus to suburban areas and to the sunbelt.

Lyndon Johnson initiated the "Great Society" with a series of social welfare enactments. In his first annual message to Congress on January 8, 1964, Johnson called for "all-out war on human poverty and unemployment . . . to help that one-fifth of all American families with incomes too small to even meet their basic needs." This campaign was designed to address the economic and social causes of poverty, rather than simply dealing with symptoms. Legislation enacted in 1964 and 1965 included the Civil Rights Act, the Economic Opportunity Act (which begat the Neighborhood Youth Corps and the Job Corps), the creation of programs offering food stamps, Medicare and Medicaid, education acts, the Equal Opportunity Act, the Voting Rights Act, and the first major low-income housing programs since the New Deal (the Housing and Urban Development Act of 1965). Signaling a shift away from filter-down approaches, the federal government exercised authority to correct social inequities, resulting in centralization of social policy-making powers.

Passed by a bipartisan majority of Congress in reaction to widespread urban rioting, the 1968 Housing and Urban Development Act marked another policy shift. Emphasizing rent supplements and home-ownership, the new law prohibited construction of high-rise public housing for families; at least half of all dwellings built with government assistance were earmarked for low-income households. The results were striking: by 1970 subsidized housing represented 29.3 per cent of all housing starts – the highest level in history – and total production reached a peak of 430,000 federally assisted units. This record was short lived, however. Only one-third of the appropriations requested by the Department of Housing and Urban Development (HUD) under the Act were funded and, in 1973, President Nixon instituted a construction moratorium.

During the 1968 campaign, Nixon repudiated the War on Poverty: "For the past five years we have been deluged by government programs for the unemployed, programs for cities, programs for the poor, and we have reaped from these programs an ugly harvest of frustration, violence and failure across the land." He did not, however, simply dismantle Great Society initiatives, acknowledging that government had a substantial "obligation to the working poor." Public spending on aid programs increased during Nixon's first five years, and the number of recipients continued to expand: by 1980, more than 21 million Americans received Medicaid coverage and food stamps, over 10 million participated in Aid to Families with Dependent Children (AFDC), and 4.2 million received Supplemental Security Income. Poverty rates fell from 30 per cent in 1950, to about 22 per cent in 1959, to 18 per cent in 1964, and 13 per cent in 1968 (the years of the Great Society), reaching a low of just over 11 per cent in 1972.

When Canadian reformers demanded social security, public housing, and health insurance in the post-war years, the government balked, declaring these measures "dangerously socialistic." Housing legislation, which eventually passed in the late 1940s, only provided for some of the working poor and certain members of the middle class. It is ironic, but perhaps predictable, that throughout most of the twentieth century, Canadian housing policies – analogous to the American pattern – were designed to assist those who were already relatively well-off, while ignoring those most in need. Housing policy was a vote-getting device, a job-creating mechanism, and a means of placating critics of low-rent publicly assisted shelter.

An amendment to the National Housing Act in 1949 provided for construction of public housing for low-income families, disabled persons, and senior citizens. In 1954 the federal government began insuring mortgage loans to facilitate home-ownership. CMHC was authorized to provide loans to municipal and private non-profit corporations in 1964. Starting in 1973, legislation introduced the notion of "income mix" to avoid large public housing ghettos. Recently, local governments have taken more initiatives to address the shelter and income security needs of low-income residents. It was not until the mid-1980s, however, that municipal officials acknowledged a growing homelessness problem.

BRITAIN'S HOUSING (HOMELESS PERSONS) ACT OF 1977

In the post-war decades homeless people in Britain could not press their grievances against local authorities with any realistic hope of success. But grassroots advocacy organizations, notably Shelter and CHAR, were established in the mid-1960s. Their formation coincided with the nationwide showing of the television film, *Cathy Come Home*, which galvanized public sympathy for homeless individuals. Pressure groups exposed both the harsh measures used by local authorities to exclude homeless people and the punitive conditions in night shelters. Voluntary organizations were not yet sufficiently experienced or sophisticated to devise policy alternatives or legislation. Instead they pointed to damning evidence unearthed by their researchers. In the mid-1970s, these voluntary agencies created the Joint Charities Group to generate media coverage of government's failure to confront the problem of homelessness. Their evidence demonstrated that legislation was necessary to bring local authorities into conformity.

Representatives of the Joint Charities Group established a rapport with key members of the Department of Environment (DoE). Their research convinced officials that, as the Secretary of State concluded, "three quarters of all authorities in England appear to take a rather limited view of their responsibilities." Nick Raynsford of the Joint Charities Group observed that

> Throughout this period the significance of the role of the voluntary agencies in filling the political vacuum cannot be overemphasized. For in the absence of any close interest on the part of ministers in the shape and detail of the legislation, and of firm and sustained political pressure for it, it would have been far more difficult without the Joint Charities Group for the DoE to resist local authority pressure to restrict the scope and effect of the Bill.
>
> (Raynsford 1986: 43)

Opposition to the proposed legislation on homelessness – a Private Member's bill – was well-orchestrated. Resurrecting old arguments that the deserving poor should be differentiated from undeserving rent dodgers, Conservative members of Parliament claimed that this "charter for scroungers" would cause local families on council waiting lists to lose out if homeless households were given priority. Public resources, they believed, were insufficient to accommodate all those unable to secure adequate housing. The legislation which passed on the final day of the 1977 Parliamentary session inevitably contained compromises. Supporters agreed to these dilutions because they felt that the government was about to change and that the Bill would not pass if reformers resisted changes. The amendments gave more discretionary power to local authorities and reduced their obligations to people who were judged to be intentionally homeless. These changes allowed for the continuation of traditional distinctions between the deserving and undeserving poor.

Nevertheless, the Housing (Homeless Persons) Act of 1977 broadened the scope of local authorities' responsibilities. It defined homeless persons as "those without accommodation they were entitled to occupy," and specifically included those (like battered women) who were "threatened" with homelessness. Priority groups were defined: they included families with dependent children, pregnant women, those made homeless by fire or other emergency, and individuals who were vulnerable through old age or mental or physical disability. The local authority was obliged to offer temporary refuge to high priority households while the facts of the case were investigated. Those judged to be genuinely homeless were entitled to council housing.

From its inception, the Homeless Persons Act has been subject to widely different interpretations by local authorities. Some, for example, regard any situations involving family disputes or rent arrears – among the principal precipitants of homelessness – as outside the scope of the law because they represent evidence of intentionality. Ironically, people's chances of being rehoused under the Act depended more on the community where they become homeless than on the circumstances or reasons for their plight. Much depended on the political will of the local authority and on the relationship between homeless people and their social worker. Some councils issued travel vouchers to induce needy people to leave their jurisdictions. A few boroughs began to use bed and breakfast hotels for temporary lodging.

Shortly after the Act was implemented a *Code of Guidance* (Department of Environment 1977) was issued, substantially broadening the Act's application. The DoE *Code* urged local authorities to take a wide and flexible view when assessing disability or poor health among applicants. It placed the onus on local officials to prove intentional homelessness. This *Code*, according to Donnison and Ungerson, "amounts to a public assertion of moral principles, as important in its field as comparable statements about race relations and equality of opportunity between the sexes" (Donnison and Ungerson 1982: 277). The *Code* does not, however, have the force of law. It has been subjected to endless rounds of interpretation by the courts. Between 1981 and 1986 there were 274 successful applications for judicial review of the decisions rendered by local authorities under the Act. About half of the High Court decisions were won by homeless applicants.

The Act was amended in 1985 and again in 1987 when the definition of homelessness was broadened. In spite of such changes, local authorities found it increasingly difficult, in the face of budget cutbacks, to live up to the *Code*'s expectations. Only half of those applying for assistance have been accepted by local authorities for rehousing. The experience of applicants varies widely, depending on the resources and willingness of local officials to shelter them: rates of acceptance in 1987 ranged from 11 per cent in Barnsley to 71 per cent in Coventry and 98 per cent in Plymouth (Thompson 1988: 21). Despite its shortcomings the 1977 Act had an immediate impact. The total number of households accepted as being homeless rose from 33,000 in 1976, to 53,100 in 1978, the first year after passage. By 1992 this number had grown to more than 184,000 households and most local authority housing departments had accepted the responsibility to deal with homelessness (Department of Environment, 1992 H1 returns).

HOUSING AND HOMELESSNESS IN THE UNITED STATES

Ronald Reagan entered the White House in 1980, having defeated Jimmy Carter on a platform which included reversing the social welfare policies of his Democratic predecessors. Determined to reduce government spending and regulation of the economy, stem inflation, and lower taxes, Reagan cut taxes for the wealthy and halted inflation by raising interest rates as the country entered a major recession. He initiated a flood of deregulation schemes and reduced funding for means-tested social programs, requesting the largest budget cuts in subsidized housing.

During the 1970s, until Reagan's election, federally subsidized housing (all programs) added 160,000 units per year to available rental housing, though many of these dwellings were earmarked specifically for elderly people. This production level declined by two-thirds during the 1980s, at the same time as the stock of low-cost housing was reduced and the number of low-income renters increased. The inventory of low-rent publicly assisted units under construction dropped from 126,800 in 1970, to 20,900 in 1980, to 9,700 in 1988. New construction authorized by HUD (U.S. Department of Housing and Urban Development) was limited to elderly and disabled persons (U.S. Bureau of the Census 1991: Table 1295). Because of demolitions, closures, and conversions, the available stock of public housing shrunk. Meanwhile, tenants doubled up and waiting lists grew: in New York City, 200,000 households are on the list for 190,000 occupied units, although the annual turnover is only 1 per cent. The number seeking public housing in Miami is 60,000; in Chicago, 44,000; in Philadelphia, 23,000; and in Washington, D.C., 13,000.

The Annual Survey of Housing conducted by the U.S. Bureau of the Census (1982) reported that the number of households with two or more related families sharing space increased by 58 per cent from the previous census, from 1.2 million to 1.9 million units.

In 1984 HUD found that 23 million households lived in substandard housing or paid too much for shelter (relative to their incomes); 83 per cent of the poor were unserved by subsidized housing programs. Among families with less than 10 per cent of the national median income, 94 per cent were unserved (Slessarev 1988: 357–379).

In keeping with the doctrine of supply-side economics, the federal government under Reagan and Bush concentrated its low-income housing efforts on the Section 8 voucher system. Eligible households were to locate rental housing which met certain standards and cost no more than a specified fair market rent. Experience was not positive for most families. The majority of qualified households was unable to find affordable housing within the two-month deadline (Riordan 1987).

Through the Economic Recovery and Tax Act of 1981, the government sought to encourage private investment in low-income rental housing (by allowing tax shelter status and accelerated depreciation). It is not clear, however, how many units were added to net stock as a result. These efforts were overshadowed by the possibility of up to 700,000 subsidized rental units being converted, and tenants being evicted, when federal use restrictions start to expire during the 1990s (Apgar 1989: 24).

In 1984 HUD issued a highly publicized report on emergency shelters. It asserted that homelessness was a temporary phenomenon, was unrelated to the severe recession of the early 1980s, and did not affect many formerly middle-class Americans (U.S. Department of Housing and Urban Development 1984). By the end of the 1980s, over 11 million people were paying more than one-third of their incomes for rent and 5 million were spending more than 50 per cent of their income on rental housing. A shortage of 4.1 million units existed for low-income renters (Appelbaum 1989).

While government agencies minimized the growing problem of homelessness, the voices of advocates grew in intensity. These groups seek to define homelessness and to hold public agencies accountable for the provision of shelter and benefits. In 1979 the Coalition for the Homeless brought a class action law suit against the State of New York and New York City. *Callahan v. Carey* asserted that shelter is a basic right which obliges the state to provide housing (*N.Y. Sup. Ct. Dec. 5, 1979*; Hayes 1989: 21). The court held that the state constitution provided for the "aid, care, and support of the needy." In 1981 a consent decree was agreed to; it set standards for the quality of public shelters, including limits on capacity, ratios of staff to residents, the size of beds, and the nature of facilities provided (e.g. laundry, mail, and telephone).

The National Coalition for the Homeless was formed to emphasize the rights of homeless people and the need for national policies focused on structural and policy shortcomings. A parallel movement, organized by the Community for Creative Non-Violence, opened shelters for homeless people and sought to publicize their belief that homelessness is not a matter of choice. Activist Mitch Snyder fasted to generate press coverage and to coerce the federal government into providing aid for shelters in Washington, D.C. These efforts had the effect of building prestige, leverage, and potency for advocacy groups, enabling them to secure funding and ultimately to affect policy. A gauge of their effect is the number of Congressional hearings on homelessness: the first hearing was held in December, 1982; in 1985 there were fifteen hearings, and in 1989, thirty-five.

SUMMARY

Many of today's public attitudes and policies toward homeless persons in Britain are rooted in a Poor Law mentality which caused them to be treated as vagrants. Government officials were reluctant to take responsibility for homeless households. It was only when social equanimity was threatened that substantive steps were taken. Even then, housing programs were generally intended to benefit those members of the working class who were reasonably well-off and represented a potent political force. Homelessness was treated as a welfare issue rather than a housing problem. Ultimately the concept of housing "as of right" was acknowledged and legislation to deal with homelessness was enacted in 1977. Implementation of the Act, however, varied widely from place to place. Repeated attempts were made in the 1980s and early 1990s to turn back the clock. Nevertheless, a number of innovative programs have been created by local authorities and voluntary agencies to cope with homelessness.

In the United States the roots of today's homelessness lie deep in American history. A punitive differentiation between deserving and undeserving poor, a concentration on programs benefiting the middle class, and an aversion to both indiscriminate charity and government intervention, all stem from established notions of individualism. Given the long-accepted penchant for privatism, a belief in the need for social control, and a conviction that government should not force social change, current American attitudes toward homeless individuals are predictable. Initiatives to provide publicly assisted housing and other aid to homeless people have not enjoyed widespread support in the United States for several reasons: social welfare programs are means-tested and thus benefit only a small group of low-income and minority Americans who do not represent a vital political constituency; housing provision and other programs directed at low-income and homeless people are more costly per household – and considerably more controversial – than such initiatives as Medicare; these social welfare schemes inevitably become intertwined with racial issues and, when coupled with civil rights requirements, are seen as threatening the autonomy of local political, social, and economic institutions.

During the Depression in Canada, when homelessness was widespread, public agencies were not eager to deal with this dilemma, nor were they well equipped to help. Officials reverted to the habit of dealing with urban problems by removing people to sparsely settled regions. Relief was spurned by individuals and by government; when help was needed municipal agencies were virtually bankrupt and unable to respond. Eventually a comprehensive social services system was assembled; it still has gaps, however. While homelessness is not a new phenomenon in Canada, both urban and rural districts have recently experienced a substantial increase in the numbers of those without adequate shelter. In particular regions housing dilemmas are exacerbated by high unemployment and reliance on public welfare or unemployment insurance. Increasing numbers of homeless people are seen on the streets in prosperous cities as well.

4

MIGRANTS AND GATE-KEEPERS[1]

The links between immigration and homelessness in Western Europe

Eliminate unemployment. Stop immigration!

(Slogan of Germany's *Republikaner Partei*)

In order to contextualize the British case study and to broaden the comparison with North America, I will take a brief look at Western Europe, where major cities are experiencing an increase in homelessness accompanied by xenophobic reactions to migrants and refugees. As rightist gate-keepers gain ascendancy, legal and institutional barriers are being raised, creating enormous disparities between residents with full rights of citizenship and a marginalized class of aliens compelled to work on the periphery within a shadow economy. Questions of race, ethnicity, gender, and cultural tolerance are at the heart of European debates over housing and homelessness, social programs, and budget cutbacks. A number of brief cases are presented from Britain, Denmark, Sweden, France, the Netherlands, Belgium, Germany, Ireland, Italy, and Spain. Though homelessness is a problem in all of these countries, it is not widely recognized as part of the national urban agenda. The number of people without shelter continues to grow in response to geopolitical and social changes, globalization, and rising unemployment and poverty. Neither the public nor private sector has been able to keep pace. Governments have succumbed to fiscal pressures, shifting responsibility for homelessness to local authorities and to voluntary agencies. Frequently, however, this devolution has resulted in neglect (in the case of municipalities which equate homelessness with unwanted foreigners) or dependence on a dated charity model of care (in the case of some religious and voluntary organizations).

In 1991 at the European Community (EC) summit Prime Minister John Major spoke ominously of the "rising tide" of immigrants, warning that "we must not be wide open to all-comers just because Rome, Paris and London are more attractive than Bombay or Algiers" (*The Independent*, June 29, 1991, London). A week earlier large numbers of Albanians were deported from Italy, and the French government forced illegal migrants to return to North Africa. The *Guardian* (June 29, 1991: 7) reported that "Britain has so far refused to sign the (Schengen) accord, arguing that there are few gains and only the enhanced risk of encouraging cross-border terrorism, drugs, crime and illegal immigration."

Two different, but related, trends are evident in the EC. On the one hand, politicians seek to gain votes by enhancing the rights of citizenship for some residents ("guest workers") who migrated long ago. On the other, gates are being closed at ports of entry; recent arrivals are treated as a shadow class of aliens without the legal rights accorded to citizens, and they are likely to suffer from racism, disaffiliation, and homelessness (Mason and Jewson 1992: 111). Although many Western European cities now house a number of ethnic groups visibly different from natives, public policies generally do not address issues associated with multi-culturalism. In its Poverty 3 Programme the EC directed attention to the political and social exclusion of 16 million migrants and refugees who have no voice in shaping the new Europe. The EC characterized these people as

> the least privileged groups of people . . . There is a considerable risk of two different societies developing within member states, one of them active, well-paid, well-protected socially and with an employment conditioned structure, the other poor, deprived of rights and devalued by inactivity.
>
> (Commission of the European Community 1990)

Reflected in a shift in emphasis from social rights to concern about the economic costs of welfare programs, this issue of social equity is linked with homelessness, which is perceived as an economic and immigration problem, and is related to the rising popularity of right-wing political groups. Among them are France's *Front National*, the Italian Social Movement, Germany's *Republikaner Partei*, the *Vlaams Blok* in Belgium, Switzerland's *Autopartei*, the *Ny Demokrati* in Sweden, Denmark's Progress Party, and the Freedom Party in Austria, which won 22.6 per cent of the vote in October 1994. These groups do not simply appeal to xenophobic or racist instincts, however; they receive backing from people who are disenchanted with politics as usual, are unhappy with the welfare state, and fear losing national identity to the European Union (Betz 1993: 413–427). They also have succeeded in pulling mainstream parties rightward, as they attempt to capture a share of the anti-immigrant vote.

Immigrant workers and their families are highly concentrated in large urban areas and, typically, within a few districts of the city or the urban fringe. Characterized as a "marginalized and alienated minority existing physically within, but economically, socially and culturally outside, affluent Western European society" this migrant population is subject to racism, segregation, and relegation to the worst rental housing (Kleinman 1992: 47; Mangen 1992: 65).

Attitudes among Europeans toward migrants prior to 1970 varied from indifference to official tolerance. Since the mid-1970s, however, most countries have taken similar steps: amnesties (for long-time residents), deportations (for new arrivals), and limitations on legal rights. The harshest measures are directed at those who are most different from citizens in terms of religious, ethnic, or racial background. A sorting process is under way to differentiate between excluded and included, to define who must go, who may stay, and what social, legal, and political rights they possess (Layton-Henry 1990). Asylum applications to EC countries increased by a factor of six between 1983 and 1991, in large part because there were no alternative modes of legal entry. However, acceptance rates continue to decline; in 1991 these rates varied from 7 per cent in Germany (which had the highest number of applicants), to 10 per cent in the Netherlands, 20 per cent in France, and 21 per cent in Britain (Freeman 1992: 1167). Since 1991 petitions for asylum have increased, in part because of the conflicts in Bosnia-Herzegovina.

During the 1980s and early 1990s Western Europe experienced de-industrialization, declining employment (particularly in manufacturing), disenchantment with large, government-sponsored housing estates, reduced public and private investment in new dwellings, a greater reliance on market mechanisms for housing provision, lower per capita social benefits, deinstitutionalization, and rising numbers of people without shelter. Accompanying these shifts has been a growing concern with, and sometimes violent reactions to, the presence of foreigners.

European housing markets are struggling to keep pace with rapid social, economic, and geopolitical changes. Population growth has been surpassed by a rise in the number of households, including single parents; the population is aging and one in four is over age 60. Gross domestic product continues to expand (by over 50 per cent in the EC between 1970 and 1989) but this trend has been offset by increasing unemployment (16 million in 1990) and poverty (approximately 45 million in the mid-1990s); these trends, in turn, are reflected in a tightening supply of low-rent dwellings and growing affordability problems.

Most states are retreating from an emphasis on social housing. Approaches are not uniform, however. They span a broad spectrum, from an attenuated form of social housing provision in Scandinavia to a trend toward privatization in the rest of Western Europe. A number of problems are also shared by these countries: housing is expensive relative to incomes, rents are rising more quickly than incomes, and debt burdens for poor people are growing, as is homelessness. More social housing stock is needed, and that which is built is not always occupied by the population most in need.

BRITAIN

Beveridge's plan was that the British welfare state would provide benefits differentially, based on traditional notions of class, race, and gender. It was assumed, for example, that women, as wives and mothers, would always live with a male provider. This view of the world did not apprehend the possibility of single parents or of women seeking refuge from domestic violence. Citizens were also distinguished from aliens; welfare citizenship depended on nationality, which was interpreted differentially by race. When Caribbean blacks came to Britain in the post-war decades to fill jobs as manual workers, they were prohibited (because of residency requirements) from moving into council housing. The education system also perpetuated class, racial, and gender inequalities (Williams 1993: 77). Politicians in the 1980s and 1990s persisted in some of these traditional notions, insisting on preservation of "family values." The system which developed (and the way in which it was administered at the local level) was ill-equipped to deal with rising numbers of immigrants, lone parents, divorced or separated women.

Ethnic minorities comprise about 6 per cent of the British population; but they are disproportionately represented in large cities, council housing, and among unemployed and homeless households. In recent years the unemployment rate for women from ethnic minorities was half again as high as the rate for white women; the rate for minority men was two to three times higher than the level for white men (*Employment Gazette*, Department of Environment, London, April, 1990). Over 60 per cent of female Caribbean household heads are in council housing, often in the poorest projects. The Commission for Racial Equality reported that "they are also disproportionately located in the least desirable sections of this housing . . . There has been substantial evidence of discrimination stretching over three decades" (Commission for Racial Equality 1984). A 1993 study found that single, female-headed Caribbean households are allocated flats rather than houses and are in high-rise buildings; moreover, "they are concentrated on higher floors or in basements to a greater extent than similar white single-parent mothers" (Peach and Byron 1993: 423).

The London Housing Unit found that blacks and other minority households were up to four times more likely to become homeless than other households. In 1991 40 per cent of hostel residents were women and nearly two-thirds of them were under age 24 and black. Sixty per cent of homeless black women are living with relatives or friends (compared to 42 per cent for whites); this doubling up reflects the existence of extended families as well as the difficulty young black women have of being accommodated in council housing (Dhillon-Kashyap 1994: 110). The situation is even worse in private rented housing. The Commission for Racial Equality found that more than one in five of the agencies surveyed consistently discriminated against blacks and Asians (Commission for Racial Equality 1990).

Within Greater London, 47 per cent of people accepted as homeless by local authorities are from ethnic minorities (*Housing Association Weekly*, February 9, 1994). Blacks and ethnic minorities being accommodated by housing associations are twice as likely to be homeless as are other new tenants (*Fact File*, The Housing Corporation's Statistical Bulletin, June 1994, No. 1, Figure 9). Ethnic minority households are twice as likely as others to live in housing which is unfit or requires major repairs.

One of the salient trends of the 1980s in Britain was a determined effort by a number of local authorities to develop anti-racist policies. These drew the ire of central government and may have speeded the demise of such agencies as the Greater London Council. The initiatives did, however, generate attention and may eventually change how the issue of racism is understood and framed in public discourse. For the moment, though, fair housing laws, fair lending legislation, freedom of information, and affirmative action policies are not as advanced in Britain as in North America; thus the actions and appeals which women and ethnic minorities can take are still quite limited.

DENMARK

During the 1960s and 1970s Denmark, one of the most homogeneous societies in Europe, experienced substantial in-migration of foreign workers, whose culture, language, religion, and physical appearance are considerably different from Danes. Initially welcomed because they filled menial jobs, the newcomers found that their reception cooled markedly during recent recessions. Increasing unemployment, lack of reasonably priced housing, and social welfare concerns caused resentment. Migrants have high social visibility, are concentrated spatially – 80 per cent live in twelve cities – and they tend to work in only a small number of industries and services. Though most Danes have no contact with the foreigners, a majority feel that too many have been admitted. Residents characterize the newcomers as dark-skinned, having lots of children, eating peculiar food, and living in apartment buildings which reek of foreign, spicy cooking (Enoch 1992: 282–300).

Residence and work permits were denied to foreigners from the end of 1970; this ban was made permanent in 1973. In 1986 the "Law for Foreigners" (passed in 1983) was tightened with respect to asylum seekers; only refugees from countries where there is an obvious danger of persecution will be considered for entry. Changes came about because of grassroots pressure, media reports of threats to Danish culture, and because of a desire for harmonization within the EC. With the exception of citizens of Nordic and EC countries (who have free access) immigration is now strictly controlled.

In January 1990 there were over 150,000 foreigners in Denmark; but only 1.9 per cent of the population were not from the EC, Scandinavia, or North America. There are now second- and third-generation residents in the migrant category. Generally unskilled and employed in trades which are vulnerable to recessionary declines, their unemployment rates are extremely high. In 1989 the jobless level for Danes was 12 per cent; but for Turks it was 44.3 per cent, for Pakistanis 41.2 per cent, and for Yugoslavs 27.5 per cent (Hjarno 1992: 75).

Homelessness is increasing, in tandem with rising unemployment, housing shortages, and tightening controls on welfare benefits. Denmark's 1974 Social Security Act obliged the counties to operate centers for homeless people. Local government is responsible for half of the social housing allocations and for subsidizing non-profit housing, which comprises about one-third of the country's total inventory. The government requires migrants to learn Danish in order to qualify for social benefits; public policy also encourages integration of migrants and "geographical spreading" of foreigners throughout the country.

Responsibility for dealing with migrants and for dispensing social benefits rests with municipalities, but there are no implementation or monitoring mechanisms in place, so local officials are left to their own devices. Using a policy guideline of "only one migrant family per staircase," some authorities have denied social housing to newcomers from Third World countries. In 1991 the courts found this practice illegal. Housing policy remains unclear, however, as the central government has not been actively involved (Hamburger 1992: 310).

SWEDEN

National and local governments in Sweden share responsibility for housing, setting guidelines, and providing financial incentives for construction and rehabilitation. Most of the country's flats are in public housing owned by municipal non-profit companies. Government renewal schemes have solved some problems and created others. About 10,000 flats per year were rehabilitated, starting in the late 1970s. Displaced households cannot afford the rents in the rehabilitated units, which often consist of "merged" flats with more space and higher rents. As a result, low-income foreigners are segregated, representing a majority of residents in poorer quality projects or districts, particularly on the urban fringe (Viden 1985).

Under the provisions of the Social Service Act local social authorities deal with people in need. They employ "social housing workers" to consult with tenants and to manage social housing, which is tied to welfare, counseling, and care, all managed by the same social authority. Acting as gate-keepers and wardens, they use housing and individual social contracts as motivators to encourage difficult clients to abide by rules and to pay rent on time. Many of these tenants have been deinstitutionalized and are placed in "training flats," intended as a transitional phase from institutional care. If these methods fail, tenants are removed. Evictions, which are more common in the public than in the private sector, increased from 4,860 in 1982, to 7,786 in 1993. A number of critics have raised objections to the obligatory linking of housing and rehabilitation programs, as well as to the high rate of evictions.

Homeless people excluded from the private market find their way to the local social authority. Along with tenants evicted from public housing, they are normally accommodated in night shelters, considered the housing of last resort. Some, however, become absolutely homeless because they have difficulty adjusting to the regimen of the hostel, or because they opt for sleeping rough. If they reject an offer of shelter, the local social authority considers them homeless by choice and does not provide further assistance. Thus, there are people homeless, sleeping rough, or in night shelters, though Sweden is a relatively wealthy country with an adequate supply of empty flats (Sahlin, 1994: 14–15).

FRANCE

Immigration, marginalization, poverty, and homelessness are closely related in France. Declaring a universal right to housing, the *Loi Besson* (1990) requires regional governments to create housing plans (though not necessarily actual dwellings) for low-income residents. While building affordable social housing for elderly and disabled persons and providing supplemental funding through the poverty program, the government devolves responsibility for pressing problems, like the need for emergency shelter, to voluntary organizations. Both sectors rely heavily on the existence of kinship networks.

French housing policy, perhaps unintentionally, reflects the public agenda for more than shelter; it deals as well with issues of immigration and social integration. It is designed "on the one hand, to keep what is non-French at arm's length from French society; on the other, to render it invisible, to dilute any possible contamination" (Lloyd and Waters 1991: 58). The available supply of low-rent housing (*HLM: habitation à loyer modéré*) is extremely limited, amounting to only one in twelve units in Paris. Approximately 2.5 million households in France encounter significant problems in paying for rent and other housing expenses. The lowest socio-economic quintile, where immigrants are concentrated, must find accommodation in the private sector. A survey by the *Conseil Economique et Sociale* found that 40 per cent of dwellings without running water are occupied by the families of immigrant workers. When immigrants do obtain access to *HLM* they are segregated; more than 80 per cent of the tenants of the Bosquets housing estate in Montfermeil, with 1,590 flats, are immigrants.

Foreign workers were brought into the country to meet industry needs during the 1960s boom. France now has 3.5 million residents who are not citizens. By the mid-1980s 2 million *maghrebins* (from Algeria, Morocco, and Tunisia) were resident in French urban areas: more than half live in just three cities: Paris, Lyon, and Marseilles. Many are single men from the Maghreb and from West Africa (guest workers) who live in overcrowded, marginal hotels (*foyers*) with inadequate facilities, owned by *Sonacotra (Société Nationale de Construction de Logement pour les Travailleurs)*. A national agency established in 1956 to shelter workers, it receives funding from public and private groups and manages 340 projects with 71,000 single rooms and 2,000 family lodgings (*Libération* 10 June 1989, Paris).

Migrants have little political power and are not well-served by public agencies. In a number of towns public officials established a limit on the concentration of foreigners (*le seuil de tolérance*), usually in the range of 10–15 per cent of the local population. Recent changes by the conservatives include tighter citizenship laws, spot identity checks, deportation of illegal immigrants, and limitations on family reunification for foreign residents of France. Immigrants who lose their jobs may now lose their residency permits as well. In 1993, the country's interior minister, Charles Pasqua, called for "zero immigration," while acknowledging that the *Front National* and mainstream conservatives "shared many of the same values." The number of deportations, 12,000 in 1994, was double the level of 1993, and another 68,000 immigrants were turned away at the borders.

In response, immigrants have resorted to rent strikes and organized squatting in order to obtain affordable housing. In some cases they occupied high-profile sites, refusing to move until their grievances were heard. In the 11th *arrondissement* of Paris, African immigrants squatted in new public housing. The municipality shut off services for several months but was unable to dislodge them. During the last decade, France has experienced political actions initiated by second generation *maghrebins* reacting against social marginalization, the rise of right-wing anti-immigration parties, police harassment, and racist killings. Riots and racial violence occurred on large, high-rise housing estates, largely inhabited by immigrants (Poinsot 1993: 92).

The anti-foreign vote – the *Front National* received 12.4 per cent of ballots cast in the parliamentary elections of 1993 – is symptomatic of social malaise.

Immigrants serve as scapegoats for increasing societal frustration with economic, ideological, and religious change. The two cultures do not mix well. Conflicts are deep, reflecting rifts over religion, socio-economic status, gender roles, literacy issues, and views on marriage. In several cities school authorities refuse to accommodate Muslim girls wearing traditional head scarves. Continued confrontation is inevitable (Todd 1991: 172).

French law requires that local authorities provide emergency shelter to anyone living in the street. Conditions in such infamous institutions as *Maison de Nanterre* in Paris, however, are so unpalatable that many people opt for sleeping rough. As a result, there are unoccupied cots in some Paris emergency shelters, while people bed down in the Metro, in cars, and in abandoned buildings.

As in other Western societies, the number and diversity of homeless people in France increased during the 1980s. Traditionally, public programs assisted nuclear families with temporary problems; today's homeless persons, however, are often young people, many of them substance abusers, who lack the skills and training to adapt to the changing job market. One in six jobless persons is doubled up with friends or relatives because they cannot qualify for social housing. One in five homeless persons is a recent immigrant. Ten per cent are women – single mothers, average age 27, with young children, half of whom receive state welfare assistance. National poverty programs began to shift in 1985 in an attempt to recognize some of these changes, but governments continue to rely heavily on voluntary agencies, which subsist on private donations and on supplemental state funding. These third sector organizations provide hostels and emergency shelters, including a barge in the Seine (*Armée du Salut*), food banks (*Banque alimentaires*), social housing (*Emmaus*), and social reintegration projects (*Aide aux Personnes Sans Abri*). One group coordinates political action campaigns led by homeless people (*A.T.D. Quart Monde*) (Spaull 1992: 39–54).

Statistics on homelessness in France are elusive; data are collected only for those people who qualify for social housing. A street count in December, 1991 found that there were 1,800 young *SDF (Sans domicile fixe)* in the Paris Metro; officials estimate that there are 15,000 *SDF* in Paris, 30,000 in the Paris region, and at least 200,000 across the country. A survey cited by Ferrand-Bechmann found that half of those seeking emergency shelter were previously doubled up with family or friends or had been evicted; the remaining half were either recently deinstitutionalized or living in the streets. Providers of free meals (*restaurants du coeur*) found that, of the approximately 400,000 users of their services, 33 per cent are single people, 20 per cent are single parents, and 18 per cent are families with three or more children (Ferrand-Bechmann 1988: 147–155).

To date homelessness has not been widely accepted as part of the national urban agenda in France. Critics argue that this issue must be recognized by including all homeless individuals in official statistics – which is not yet the case – and by allowing for broader, more inclusive definitions in order for needy people to qualify for social housing.

THE NETHERLANDS

The Dutch regard housing as a social good. Public intervention is generally accepted and most units receive some form of subsidy. Problems are apparent with post-war housing, however, particularly on large social housing estates. These include poor housing quality, high rent burdens – one in four social housing tenants is in arrears – rapid tenant turnover, a growing need for maintenance due to vandalism and technical defects (particularly in systems-built post-war projects), and declining neighborhoods.

In the early 1990s the Netherlands had 850,000 immigrants (6 per cent of total population), including 210,000 Surinamese, 177,000 Turks, 140,000 Moroccans, and 66,000 Antilleans. These foreigners are highly concentrated: half are located in the four largest cities, Amsterdam, Rotterdam, The Hague, and Utrecht, where they represent 13.4 per cent of the population. Most live in inner cities where immigrants and their children sometimes make up half of the population at the district level.

As a result of globalization the nature of jobs in the Netherlands is changing, and unskilled immigrants are the first to be made redundant. In addition to being segregated in the least desirable districts, and being assigned to the worst housing projects, they also have the least education and the highest unemployment rates. At the beginning of the 1990s, when unemployment among native Dutch workers was 9 per cent, the rates for other groups were significantly higher: Moroccans, 43 per cent; Turks, 39 per cent; Antilleans, 32 per cent; Surinamese, 31 per cent; Italians, Spaniards, Portuguese, and Yugoslavs, 16 per cent (Muus 1991: 29). Recent surveys report that, not only are migrants highly concentrated in segregated urban districts, but there is little upward social mobility and labor market prospects are bleak: "Such poor prospects are caused less by the characteristics of the minority labour force than by the nature of the labour market's selection processes, including the discriminatory practices of employers" (Roelandt and Veenman 1993: 141).

Government policy in the post-war era was to encourage immigration in order to fill menial jobs considered undesirable by Dutch citizens; officials believed that newcomers would return home after earning a "nest-egg." Assisted repatriation schemes were introduced and welfare groups were subsidized to help immigrants return to their birthplaces. By the end of the 1970s, many immigrants were unemployed, but it was apparent that repatriation was not working. In 1986 legal immigrants won the right to vote in local elections. Public policy is now aimed at promoting integration and racial harmony. Non-white politicians recently have been elected in a number of communities.

BELGIUM

Belgium is a prosperous country with an extensive welfare state. Since 1919 the *Société Nationale du Logement* has provided dwellings and housing subsidies to low-income families. During the post-war decades from 40 to 50 per cent of all new residential buildings were subsidized (Mabardi 1985: 75). Starting in the 1980s, the government provided reception centers for homeless people

31 and *32* Interior and exterior photographs of a system-built public housing project in Middleburg, the Netherlands, built in 1971–1972. Residents are concerned about their safety because gallery-access units allow strangers to roam the hallways. The sign of a guard-dog warns: "I'm watching here! If you enter it is at your own risk." Because of design flaws and social problems the local authority decided, in the mid-1980s, to dismantle the top seven floors and to reassemble them elsewhere as three-storey and four-storey buildings (Photographs by author 1985)

and destitute migrants. In the post-war era Belgium depended on cheap, foreign labor to do jobs shunned by natives. At the beginning of the 1990s foreign workers and their families represented about 10 per cent of the population. With a changing economic structure and job market, however, and the decline of manufacturing and heavy industries, poverty and unemployment grew. Dwindling tax revenues necessitated government reductions in social and welfare benefits. Recent recessions fostered the rise of rightist parties advocating repressive measures against foreign workers. Limits were placed on immigration for family reunification in the mid-1980s, and unemployed foreign workers were offered bonuses to repatriate.

Boroughs are allowed to refuse residency to foreigners. Municipal officials argue that this is necessary "as soon as the number of migrants exceeds a certain density which endangers national and public security, the maintenance of public order, the prevention of crime, the protection of health and morality, and the rights and freedoms of others." Restrictions on foreigners from developing countries (among them former Belgian colonies), including overt segregation in housing and assignment to the worst housing estates, extend even to migrant workers' children who were born in Belgium (Merckx and Fekete 1991: 67–78).

GERMANY

Germany has had a "foreigners law" since 1939 which requires people to obtain a residence permit and a work permit before entry. This law was modified in 1990 to differentiate between migrants with unlimited residence permits and others. Unlimited residence can be secured only through naturalization, which is subject to so many conditions that fewer than 3 per cent of migrants have succeeded in becoming naturalized. The remainder are liable to expulsion. Migrants represent about 15 per cent of Germany's work force. But only 11 per cent of the migrants are from Africa or Asia; most are ethnic Germans from other parts of Europe, people from other EC countries, or people from countries like Britain, the United States, or Canada, which have favored-nation status.

In recent years there has been a resurgence of neo-fascism and racist violence, directed at boat people and newcomers from Africa or Asia. These recent arrivals from developing countries are confined to hostels and not allowed to travel or to associate with Germans. The German government has actively tried to discourage migration from Asia and Africa – German officials show overseas applicants videos of the poor conditions in hostels – and has offered repatriation premiums to foreign workers. Still, almost two-thirds of Germans think that there are too many foreigners in the country. Many feel that the jobs formerly held by foreigners now must be given to refugees from East Germany. There is a widespread belief that Germanness should be defined by biological and cultural homogeneity, and that dark-skinned Turks and other foreigners, who do not speak proper German and are practicing Muslims, should either leave the country or should be compelled to take only those dirty jobs that Germans do not want (Räthzel 1990: 38–46).

Like other Western European countries Germany needs more affordable housing. The German Institute for Economic Research (*Deutsche Institut für Wirtschaftsforschung*) reports that the country requires about 5 million new units by the year 2000 to balance supply with demand. The lack of affordable housing affects single-parent families and poor families with many children. The shortage, especially acute in Berlin, Munich, Stuttgart, Hamburg, and Cologne, is exacerbated by demographic trends: household size is declining, elderly people are living longer, often remaining in their own flats, divorces are increasing, as is the number of people released from institutions, and large cities are besieged by refugees from the East. Racial incidents have increased dramatically – well over 1,000 were recorded in Germany during 1994 by the London Institute of Race Relations – as unemployment levels rose along with large numbers of newcomers competing for jobs, housing, and social benefits. Government responded in the early 1990s by curbing its generous asylum law, reducing welfare assistance and unemployment benefits, curtailing public investment in social housing, abolishing tax relief for housing associations, and by emphasizing owner occupation – the top quintile in Germany now receives roughly the same amount of subsidy (direct and indirect) as the poorest quintile.

A profile of housing and homelessness in Hamburg

Hamburg (population 1.6 million), like other large German cities, is undergoing drastic changes resulting from globalization. Significant social costs are associated with these trends. For the lowest socio-economic groups further marginalization has been the legacy of the 1980s. The number of Hamburg residents receiving housing subsidies increased from 64,000 in 1981 to 84,000 in 1987; the rent burden for low-income residents exceeds 63 per cent of income (three times the city average); and, between 1988 and 1989, the number of homeless individuals more than tripled, from 8,000 to 26,000; the level in the mid-1990s is estimated at 43,000.

In Hamburg there are 3,700 bed spaces for men in hostels and pensions, and a further 1,000 places in small living communities. Only 230 places are specifically reserved for women, 70 of these for women with children. Move-on accommodation has traditionally come from the private sector or from one of the three housing associations linked directly with the City. However, as the City's contracts with these groups expire, the available supply of social rented housing units is diminished (Ossenbrugge 1990: 105).

The City State of Hamburg spent 25 million Marks on emergency shelters in 1993, as well as 110 million Marks on housing homeless and refugee claimants as well as ethnic Germans from Eastern Europe in cheap rooming houses. They typically pay 20–50 Marks per bed per person; despite these expenditures, complaints about standards of cleanliness are common. An association of German cities (*Der Deutsche Stadtetag*) reports that municipalities spend six times more on these expedients than if they paid rent arrears to allow people to remain in their flats, as is done now in Cologne.

In recent years large housing estates have been caught in a downward spiral, characterized by increasing vacancies, reduced subsidies, higher interest rates, and rising construction costs, leading to higher rents. This, in turn, compelled middle-income people to move out of these estates because they could find better housing at comparable rents elsewhere. As a result, more immigrants, people on social assistance, and "problem tenants" moved into the large projects, precipitating relocation by more middle-income people, causing costs to rise further. Some local authorities dealt with these problems by introducing structural changes in the operation of large estates as well as changes to the physical environment. Maximum rents were reduced and private non-profit organizations were allowed to assume control of the maintenance and rental operations. The result has been to attract more middle-income tenants, to reduce vacancy rates, and to break the downward spiral which was slowly causing the deterioration of the large housing estates (Friedrichs 1988: 89).

Social and economic changes are reflected spatially. Public housing residents are concentrated in older high-rise blocks which constitute about 80 per cent of total public stock. Increasing socio-spatial inequalities are evident, with growing segregation of the very poor. Gentrification – about 100,000 units per year – occurs in the central cities, displacing the poor to the periphery, similar to the pattern in France (Dangschat and Ossenbrugge 1990: 86–105). These districts attract low-income single parent families, the unemployed, elderly persons, drug addicts, refugees, ex-prisoners, and people released from institutions.

A sharp dichotomy is emerging: improved space and standards are available for the majority; but a growing number of low-income households are becoming homeless. Some prefer to squat on land or in empty houses or warehouses rather than go into public shelters. Accommodation provided by public and private agencies includes new shelters for people applying for asylum (*Asylbewerber*), containers, tents, and ships which are used for emergency shelter, as well as parks for people sleeping rough (Specht-Kittler 1992: 27–37). Large municipalities operate hostels but, even on the coldest nights, they have vacant spaces because homeless people fear theft and many object to the rigidity and paternalistic structure of the public shelters.

As the state's budgetary problems grew in recent years, greater reliance was placed on large voluntary groups. A major role is played by these welfare organizations, but they continue to deal with people in traditional fashion, some regarding homelessness as evidence of personality defects. Between 1985 and 1990 the number of counseling services for homeless people doubled; this work, however, is impeded by a lack of affordable housing, and the wait for "move-on" housing is now as much as one year.

IRELAND

At the end of the 1980s per capita income in Ireland for the poorest quintile was only $2,817, about the same as Spain, which is the lowest in Europe; this compares with $3,789 in Britain, $4,271 in France, $4, 441 in Italy, $4,662 in the U.S., $4,943 in Germany $5,027 in Canada, and $5,944 in Sweden (Townsend 1993: 8, Table 1.1). The country's poverty is mirrored in the faces of homeless people, whose numbers increased in the 1980s as jobs disappeared. It is common for young Irish to search for work in England and North America; but large numbers later return to Ireland, often in destitute condition, having neither money nor access to housing.

The right to a home is enshrined in the Irish Constitution and is a stated aim of Irish shelter policy. Putting policy into practice, however, has proven difficult. Benefits have accrued mainly to middle-income households while "those at the bottom end of the housing ladder have suffered an absolute decline in housing standards" (Blackwell 1988). Inequities are related to the problem of securing private rented, owner-occupied, and public housing. Barriers to entry (deposits, rent advances, high payments, and discrimination) are most significant for those with the lowest incomes. They also face problems because of poor conditions and insecure tenure in private housing and inadequate maintenance as well as overcrowding in public housing (Dillon and O'Brien 1982).

The obvious place for poor people is the public sector, but government policy was framed with the nuclear family in mind; traditionally, public authorities did not provide shelter for singles under age 65, unmarried couples, or lone parents. In recent years, as single parents have been accommodated by local authorities, they are highly segregated. In the mid-1980s, for example, 45 per cent of all lone parents housed by the Dublin Corporation were placed in one project, which contains only 10 per cent of the city's public housing stock; 60 per cent of Dublin's homeless singles who sought public accommodation were assigned to this project, which now has a reputation as being a depository for people with serious social problems. In recent years, as homelessness increased, construction of public housing came to a virtual halt; this was accompanied by a sale of the best dwellings, thus limiting the options available to lower-income groups (Kelleher 1990).

Ireland's 1989 homelessness legislation is the country's first official recognition of the needs of homeless people. It is not mandatory, however, and makes no provision for financing or implementation nor does it assign responsibility for hostel provision. Care for the poor in Ireland, especially young people, is shared by a host of public, voluntary, and religious agencies in uncoordinated fashion: these include the police, social workers, probation officers, juvenile liaison officers, voluntary agencies, workshop supervisors, and house parents at sheltered homes. Most agencies have no idea where homeless young people go after leaving; many simply drift into squatting, sleeping rough, or moving from shelter to shelter.

In most cases, responsibility for accommodating homeless people has been delegated to municipalities or voluntary agencies. Advocates are highly critical of this system, arguing that undue reliance is placed on well-meaning but untrained workers in religious and voluntary groups. With limited resources, these organizations are dependent on handouts; they devise make-work schemes that do not correspond to the job market; workers customarily neglect to consult with users, and most follow outdated, severely circumscribed programs based on a charity model.

ITALY AND SPAIN

Italy, Spain and Greece present a paradox. Characterized as "the leaky door" of the EC, they were disparaged by members of the European Union as being too lenient in admitting African migrant laborers. In the interest of harmonization, since joining the EC, they have been forced to bar the door. Migrants, however, continue to fill a crucial economic niche. Fishing boats in the south of Italy and Spain, for instance, often are crewed by low-paid Tunisians and Moroccans. Half of Italy's vital tomato crop and a major part of southern Spain's agricultural produce is picked by illegal immigrants from Africa. Many maids hired through the underground economy in Spain are "clandestines" from Africa or the Philippines who, in effect, are captive because they have no legal rights. They live under the same roof as their employers, are economically marginalized and socially isolated. While this practice helps to ensure cheap prices – as migrants' wages are low and they have no job security – it raises real concerns: Italians have the lowest birthrate in Europe, the ratio of active workers to pensioners has declined steadily, and natives are being urged by politicians to produce more babies "to keep away armadas of immigrants from the southern shores of the Mediterranean" (Labor Minister Carlo Donat Cattin, quoted in Milanesi 1992: 14).

In Italy, Spain, and Greece there is no legislation addressing homelessness. Statistics in Italy deal only with those evicted; in Greece no data on this phenomenon are gathered. As a result, it is only when there is a catastrophe, like earthquakes or floods which affect natives as well as migrants, that homeless people receive public attention. (Milanesi 1992: 16).

Spain has 800,000 non-European immigrants; roughly 300,000 are illegals. There are more than 6,000 homeless people living in 185 reception centers, and another 3,000 who use night shelters, as well as a transient population (*Transeuntes*) who move about the country to find better working conditions and housing. Spain enacted the *Ley de la Estranjera* (foreigners law) in 1986 to curb the influx of migrants from North Africa and to convince its European neighbors that it would be fit for entry into the EC in 1992. Since then, the country has witnessed an increase in the number of illegal immigrants, harassment by the police (which is permitted under the *Ley de la Estranjera*), and more evidence of popular and institutional racism. To date, however, there has not been a major national political movement to capitalize on racist and anti-immigrant sentiment, nor have immigrants been able to consolidate their grievances and form a political alliance.

Italy now has about 500,000 immigrants from developing and Eastern European countries – including recent arrivals from Bosnia-Herzegovina – accounting for roughly 1 per cent of the population. While working in areas of economic instability, many are socially isolated, suffering from poverty and homelessness. The pattern of dealing with migrants in Italy and Spain is characterized as "a long-term relaxation of the internal frontiers and a rapid reinforcement of external ones" (Sciortine 1991: 89). "Stop policies" have operated for some time but are ineffective, as they cannot stanch the flow of migrants into an unregulated shadow economy. Official actions taken to define these workers as "clandestines" are in part symbolic, to appear to be doing something in response to the shortage of jobs; partly political, to harmonize with the EC; partly administrative, to differentiate between the included and the excluded; and partly motivated by a fear of Middle Eastern terrorism. The first bill on immigration control in Italy, for example, was proposed in 1985, immediately after a terrorist attack at the airport in Rome. Since then, "illegal" has been synonymous with "unwanted." In 1988, for example, 81 per cent of the charges, and 26 per cent of the arrests against foreigners in Italy were related to their clandestine status (Sciortine 1991: 98). The last amnesty proved to be extremely popular. In a country with a bewildering array of political parties and electoral coalitions, this measure passed Parliament with an 85 per cent margin.

Another reaction has been from the right, directed at people of color who are homeless and jobless. A couple of hundred homeless Indians, organized as the United Asian Workers' Association, staged a hunger strike in Milan in 1990. They occupied a church but were removed by police. After squatting in an empty council building they were evicted because it was alleged to be unfit for human habitation. Another case involved a group of 600 homeless Moroccans who established a shanty town at *La Cascina Rosa* on a derelict farm outside Milan. The *Carabinieri* moved in at night, evicted them, and razed the town on the grounds that it was a health hazard. This was widely seen as an attempt by sitting politicians to appease the right, whose power increased along with their xenophobic appeals. It also helped to assure other members of the EC that Italy would no longer tolerate leaky borders.

In other cities – and homelessness in Italy is an urban problem – there has been a variety of responses. Turin and Verona in the industrial north banned immigrants from washing automobile mirrors and windshields at traffic lights. A hostel used by 120 homeless *Maghrebin* in Bolzano was razed by arsonists after town councillors indicated that they did not want any more Africans in their district (Kazim 1991: 84–89). In conjunction with the 1989 amnesty, tough new alien laws were enacted, providing for the expulsion of illegal immigrants lacking means of support. Visa requirements for North Africans and Turks were introduced as part of a computerized registration system (common to Western European countries) for all immigrants along with the stipulation that those who were not regularized must be deported.

Substantial differences are evident between poor native Italians and migrants. Both groups have been adversely affected by globalization. Industrial employment in Italy fell by 20 per cent between 1981 and 1991, and the rate of loss was greater than 26 per cent in Naples. Both groups have very low incomes: per capita monthly incomes are below 250,000 lire (£100), of which 60 to 70 per cent is spent on food. Poverty is high in the South (26.4 per cent) and in the peripheral public housing districts, characterized by decrepit projects and poor public services, and some inner city districts elsewhere in the country: "entire quarters that are in decay and impoverished by the absence of tenured job opportunities and exclusion from even minimally effective educational, health and welfare services." Even in the relatively affluent North:

> the particularly high cost of living, the increase in poorly paid and non-tenured jobs, the chronic housing crisis and the inefficient public welfare system within the framework of growing social heterogeneity and fragmentation are calling into question the assumption that poverty is a very limited phenomenon.
>
> (Mingione and Morlicchio 1993: 414)

What differentiates the two groups and causes more migrants to become homeless is the role of discrimination and lack of networks which newcomers can use to their economic advantage; the crucial role of kinship networks and reciprocity among native Italians enables them to maintain a low, but marginally acceptable, standard of living. These extended families are especially important because the voluntary sector is almost non-existent, and such networks are essential in providing access to the informal economy.

SUMMARY

Across Western Europe barriers are being raised, creating a marginal class of aliens working within a shadow economy. The cases from Western European cities illustrate the complex web of issues surrounding housing and the connections among globalization, immigration, marginalization, poverty, and homelessness. Questions of race, ethnicity, gender, and cultural tolerance are at the heart of the debate over social programs, budget cutbacks, and the emergence of rightist political parties.

Homelessness is increasing in large cities as a result of geopolitical and social changes, globalization, and rising unemployment and poverty. Neither the public nor private sector has been able to keep pace with these events; as a result, the supply of affordable housing is constricted, maintenance is neglected, and disparities between average citizens and those in the lowest socio-economic quintile are widening. After resorting to the use of temporary housing, containers and barges to shelter homeless people, governments have succumbed to fiscal pressure, devolving responsibility for homelessness to local authorities and to voluntary agencies. Frequently, this has resulted in neglect.

In these countries migrants are marginalized, confined to menial jobs, segregated, and relegated to the worst housing (usually private sector rentals) in the least desirable districts of large cities. A growing polarization is evident between average citizens and those without full rights. Pressure is mounting for newcomers to conform, to speak the national languages, and to integrate into mainstream society. Migrants have responded by demonstrating, squatting, and staging rent strikes. Typically, their protests are ignored. Though most are people of color, they remain invisible.

Governments have not been equal to the task of dealing with refugees, migrants, or people released from prisons, asylums, or mental hospitals, in part because this is a costly process which does not enjoy widespread political support. As a consequence, substantial numbers of those at the bottom of the socio-economic ladder are now homeless. Relevant trends for selected Western European countries are summarized in Table 6.

Table 6 Selective comparison of relevant trends in Western Europe

	Germany	*Denmark*	*France*	*Italy*	*Ireland*	*U.K.*
Legislation	Foreigners Law (1965); modified in 1990 to differentiate between long-time foreign residents and newcomers	Foreigners denied residence and work permits since 1970 unless they are genuine refugees. More restrictive regulations since 1986	Foreigners do not have full rights of citizenship. In 1990s Conservatives tightened citizenship requirements and limited family reunification for foreign residents	No legislation addressing homelessness; "Stop policies" have been enacted to restrict the entry of foreigners along with amnesty acts for long-time residents from other countries. Tough alien laws enacted in 1989; visa requirements for North Africans and Turks	Passed homelessness law in 1989; not mandatory that homeless people be housed by local authorities	The Housing (Homeless Persons) Act of 1977 gives housing priority to dependent children, pregnant women, and those vulnerable due to old age or disability. Local authorities obliged to provide housing
Migrants	15% of the workforce; older migrants from Turkey and less-developed European countries; newer arrivals from Asia and Africa as well as major influx from East Germany	Only 1.9% of population is not from EC, Scandinavia, or North America	3.5 million foreigners; about 2 million from North Africa; more than half of the foreigners live in just three cities; highly concentrated in poor housing on urban edge	500,000, equal to 1% of population; the state differentiates between the included and excluded, partly based on a fear of Middle Eastern terrorism	Virtually no foreign workers; almost all homeless people are Irish, many of them recently arrived back from seeking work in England or the United States	Half of the homeless in Greater London are ethnic minorities; ethnics four times as likely to become homeless; two-thirds of hostel residents are blacks/Asians
Economic trends	10% poverty rate. Foreigners are concentrated in jobs which are vulnerable to de-industrialization and therefore have higher unemployment rates	Migrants take unskilled work; relatively low poverty rate but unemployment is high among migrants: 44% for Turks, 41% for Pakistanis, 27% for Yugoslavs.	16% poverty rate. Immigrants, many of them single men, are highly concentrated in a few industries. Unemployment has led to increased homelessness; at least 200,000 *SDF* in France. About 2 million households have serious economic and housing problems.	18% poverty rate. Italy needs foreign workers for menial jobs because native birth rate is lowest in Europe.	19% poverty rate is highest in Europe. Improvements since joining EC. Housing conditions poor in both private and public sectors.	% of very poor people (below half average income) doubled since 1971; a third of manufacturing jobs lost during 1980s; Unemployment rates for Bangladeshis (47%), Pakistanis (30%).

Table 6 (cont.)	*Germany*	*Denmark*	*France*	*Italy*	*Ireland*	*U.K.*
Public policy	Most foreigners liable to expulsion; no right to citizenship. Rightist politicians have vowed that no more Turks will enter Germany.	Municipalities responsible for dealing with migrants. They disperse families: "only one migrant family per staircase." Counties have operated centers for homeless people since 1974	Homelessness not part of the national agenda. Responsibility shifted to kinship networks and the voluntary sector. Some migrants given repatriation bonuses to leave France	"Regularized" long-term resident foreigners are allowed to stay. Sitting politicians seek to appease the right with restrictions on aliens, especially from North Africa and Turkey; those not regularized can be deported. Albanian refugees refused entry	Homeless people concentrated in worst housing estates. Much responsibility is shifted to the voluntary sector	Homeless families in B & B and temporary housing; some assigned to problem estates of high-rise council housing
Public response	Migrants from Third World confined to hostels; two-thirds of Germans feel there are too many foreigners. Resurgence of neo-fascism and violent racism in recent years	Some resentment, official attempts to compel acculturation and geographical dispersion, but violent racism is not evident	Third sector provides food, emergency shelters, social housing, and political advocacy. Riots and racial violence on large housing estates inhabited by immigrants	Upsurge of xenophobia in 1980s and 1990s; migrant squatter settlements razed by police. Homelessness not regarded as a major issue because it primarily affects foreigners	Heavy reliance on church-based social welfare system run by well-meaning but untrained workers; punitive charity model and make-work schemes often guide the response to homeless individuals	Migrants in worst housing; some sent home to Bangladesh; highest unemployment rates are ethnic minorities
Political trends	*Republikaners* advocate cessation of immigration to save jobs. Extreme right elected in several states	No major rightist political movement	The rightist *Front National* has significant voter support	Rightist politicians calling for higher birth rate so the need for foreign labor will diminish	Homelessness not recognized as a significant political issue	Conservatives in power since 1979; movement to privatize housing and some welfare functions. Right-wing won some local by-elections

PART II

THE HUMAN DIMENSIONS OF HOMELESSNESS

What are the physical, social, and psychological implications of being homeless? This section examines the concept of home and the experience of homeless people in order to understand what it means to be without adequate or secure housing.

First, the health issues – both physical and mental – associated with homelessness are explored in Chapter 5, where the links between health and homelessness are discussed, as is the difficulty some have in obtaining proper health care and associated services.

The phenomenon of deinstitutionalization is analyzed in Chapter 6, which introduces the concept of community-based housing and social programs as an alternative to institutionalization.

Chapter 7 describes the experience of homelessness from the perspective of the people involved: single mothers and battered women, children and youth, veterans, people with disabilities, the frail elderly, refugees and new immigrants, Natives and visible minorities, and chronically homeless individuals.

Examples of commonly used but substandard shelter alternatives for homeless people are provided in Chapter 8: bed and breakfast establishments in Britain and the welfare hotels used by New York City and other municipalities. Emergency shelters are examined, using Toronto as an illustration of the inadequacies of this system and its effects on people who are vulnerable because of ill health, because they have been deinstitutionalized recently, or because they need more than a barracks-like hostel to get back on their feet.

5

A PRESCRIPTION FOR POOR HEALTH[1]

> Why do you wait until I'm sick to give me the housing and stability I need to keep me healthy?
>
> (A homeless man with AIDS in Boston)

> Among the good reasons to "do something" about homelessness is that homelessness makes people ill – it is unhealthy for children, for their parents, and for other living things. In the extreme, it is a fatal condition.
>
> (Wright 1989: 84)

Ironically, within a few hundred yards of the most advanced scientific and medical establishments in London, New York, and Toronto, thousands of homeless people suffer from a variety of health problems. In addition to economic deprivation, itself one of the most serious health hazards, these include: tuberculosis, hypertension, respiratory problems, skin ulceration, and a variety of other infectious diseases. Ill health for persons without adequate shelter is attributable in part to institutional and attitudinal barriers to the provision and delivery of health care.

Often the result as well as the cause of poor health, homelessness contributes to illness through a number of factors: the absence of a "home base" and supportive network, physical and psychological stress from exposure to the elements and from living in crowded, chaotic, unhealthy environments, lack of protection from an array of bacteria and viruses, and social problems associated with poverty and the stigma of being on the streets.

A thorough survey at health clinics in nineteen American cities found that poor physical health is cited as a factor in homelessness by one in five women, one in four men, and one in three chronically homeless people. Wright and Weber concluded that "the homeless probably harbor the largest pool of untreated disease left in American society today." Over half of those for whom poor health was the single most important factor in homelessness were "definitely not employable" (Wright and Weber 1987: 17).

Part of the Health Care for the Homeless program, these clinics were founded in December 1984 with $25 million contributed by the Robert Wood Johnson Foundation and the Pew Memorial Trust; for the moment at least, funding continues under the health care component of the Stuart B. McKinney Homeless Assistance Act. The facilities used are shelters, soup kitchens, missions, alcohol detoxification and drop-in centers, and juvenile courts. In the first full year of operation, to the end of March 1987, the clinics had 173,000 contacts with 59,000 different individuals; 64 per cent were male adults, 26 per cent adult women, and 10 per cent homeless children under the age of 16. About one-quarter of the adult women have dependent children in their care (Wright and Weber 1987: 56). Data from these clinics, and from other American, British, and Canadian studies, are cited in the following section.

HEALTH PROBLEMS ASSOCIATED WITH HOMELESSNESS

Among the health problems commonly encountered by homeless people are the following:

Cold injury In Canada, the northern United States, and Britain, homeless people are vulnerable to hypothermia and frostbite because they are "sleeping rough," they lack adequate housing, clothing, and health care. Often they are sheltered in hostels which require that they leave every morning, to search for work, regardless of weather conditions.

Cardio-respiratory diseases At Health Care for the Homeless clinics one-third of the patients have upper respiratory problems. As many as 30–40 per cent of homeless people suffer from such chronic physical illnesses as coronary artery disease, high blood pressure, or emphysema. There is also a high incidence (compared to a national sample of adult primary-care patients) of hypertension, gastrointestinal disorders, skin problems (impetigo, boils, and ulcers), injuries (lacerations, sprains, bruises, and fractures), and dental problems. Because many are constantly moving and cannot receive regular medical care their injuries and illnesses frequently become disabling.

Tuberculosis During a five-year period (1987–1992) the United States experienced an increase of 20 per cent in active tuberculosis cases. TB is highly infectious and is common among residents of shelters and those living on the streets. The rate of TB among homeless persons is at least a hundred times greater than the average for the general population. A study in American cities concluded that about one in six "has an infectious or communicable disorder that poses some potential risk to the public health" (Wright and Weber 1987: 110). These diseases spread quickly among people who frequent hostels: one man in Maine with TB infected over 400 others in the same facility. Tuberculosis is difficult to cure in a mobile population. Follow-up studies suggest that an extremely small percentage of homeless TB patients continue taking their medication after being discharged.

Skin diseases Homeless persons are predisposed to skin problems and edema resulting from malnourishment, poor circulation, ill-fitting shoes, cuts, and dirty clothing. The places where they sleep usually lack laundry facilities for them to wash their clothes. Individuals are affected by intestinal worms, scabies, and lice because of unsanitary conditions. Delousing is still a common practice at shelters (Burt and Cohen 1989: 49, 89).

Nutritional deficiencies Because of poor nutrition and lack of care, a high percentage of homeless persons have dental problems and suffer from

TB patients

They are seen in emergency rooms and diagnosed and maybe admitted for a couple of weeks. But there is no housing for them. They come back to the shelter. They quickly get off their medication, they [are] not eating properly and they cough through the night and the person sleeping next to them gets tuberculosis.

(U.S. Congress 1990: 36)

malnourishment, which increases the risk of infectious diseases and gastro-intestinal disorders. Moreover, the nutrients of greatest concern, vitamin C, thiamin, and folate, are not present in the foods provided by most soup kitchens because these places do not have either adequate refrigeration or a regular supply of fresh fruits and vegetables. People lacking these vital nutrients suffer from weakness, fatigue, depression, and emotional disturbances.

Sleep deprivation For most people sleeping is difficult in the noisy, turbulent atmosphere of emergency shelters. Sleep disorders cause irritability, apathy, or behavioral impairment. Shelter children, in particular, are adversely affected by stress and lack of sleep.

Health problems of children and youths Children in emergency shelters, bed and breakfasts (B & Bs), and welfare hotels have emotional and developmental difficulties; most are unable to do well at school. It is common for these children to move from one school to another as their family shifts from shelter to temporary housing to permanent housing, all in different parts of the city. Wright's observations of the National Health Care for the Homeless Program led him to conclude that among children in homeless families "chronic physical disorders occur at approximately twice the rate of occurrence among ambulatory children in general." About 16 per cent of the homeless children seen in the health clinics already have chronic health conditions. He found that their poor health and chronic physical illness

> contribute to the cycle of poverty . . . interfere with, if not preclude, normal labor force participation, and with it, the ability to lead an independent adult existence . . . Poor health may be one mechanism by which homelessness reproduces itself in subsequent generations.
>
> (Wright 1989: 65, 67, 72–73, 77)

The rate of chronic disease among youths is nearly twice that of their peers in the general population. They are subject to upper respiratory infections, traumas and skin disorders, lice infestations, chronic problems with eyes, ears, and teeth, along with malnutrition, gastrointestinal disorders, genito-urinary difficulties, and sexually transmitted diseases. A substantial percentage abuse drugs and alcohol.

Mental ill health There are evident links among homelessness, mental health, and public policies relating to the care of mentally ill people (Belcher and DiBlasio 1990). Though numbers are elusive, it is generally accepted that about one in three homeless people has serious and chronic forms of mental illness. Rossi reported that nearly one-half experience "hopelessness and despair concerning [their] prospects." The prevalence of depression is manifest in a significant number – one out of six – who attempt suicide (Rossi 1989: 145–156).

These problems are worsened by the stress of homelessness and institutionalization. A significant issue is the fate of these people once they have been deinstitutionalized; this topic is explored in Chapter 6. Rossi concluded "that there is still a heavy traffic back and forth between the mental hospital and the homeless condition." About one in four homeless persons in the United States has been in a mental hospital at least once. This is more than five times the rate for the general population. A Los Angeles survey found that 28 per cent were chronically mentally ill: "both substance abuse and schizophrenia were elevated among individuals who had been homeless many times or for long periods of time" (Koegel and Burnam 1988). Wright and Weber concluded (and others concur) "that about two-thirds of the homeless are either alcoholic, drug abusive or mentally ill and that nearly a quarter of them have two or more of these disorders" (Wright and Weber 1987: 94).

Physical and sexual assault Life on the streets is violent. Physical assaults and muggings are common, and these attacks precipitate health problems. Women and young people are extremely vulnerable. Homeless women frequently suffer from trauma. The rate of sexual assault in this population is twenty times higher than for women in general (Kelly 1985: 75–92).

Drug dependency Between ten and fifteen per cent of homeless males abuse drugs. Among those who do so regularly there is a high occurrence of liver disease, HIV/AIDS, venereal disease, skin ailments, bruises, lacerations, and injuries resulting from violence. The U.S. Public Health Service conducted a study of homeless youth at three primary care clinics in San Francisco. They found that 6.5 per cent have been in a drug treatment facility, 15 per cent use IV drugs, one in eight has drug-using partners, and almost one in ten indicates that drugs are a problem; between 4 and 8 per cent of these young people are HIV-positive (Sherman 1992: 433–440).

Mortality The death rate for homeless people is about four times greater than the rate for the general population. Among young homeless men the incidence is even higher. On average, homeless adults die twenty years earlier than their non-homeless counterparts. In a survey of adults at Health Care for the Homeless clinics it was discovered that over half die violently: one-quarter of those who died were murdered, a rate more than twenty times higher than the U.S. average. This was followed by deaths directly attributable to alcohol and drugs (16 per cent of deaths), car accidents (13 per cent), and by cancer and heart disease (11.5 per cent each) (Wright and Weber 1987: 126–137).

HIV/AIDS AIDS patients represent from 5 per cent to 11 per cent of all homeless people. Because drug use is common among street people, and some are gay and/or engage in homosexual or heterosexual sex to earn money, they are susceptible to HIV/AIDS and to sexually transmitted venereal diseases. A study at health clinics for homeless individuals in Boston found that 78 per cent of the patients with AIDS or HIV had a history of IV drug use; 18 per cent were women.

HIV-AIDS patients

Persons with HIV who come to us are often very ill and weak from marked weight loss, fever, shortness of breath, and daily diarrhea. They have great difficulty surviving the shelters, often having been discharged from the hospitals while still recovering from pneumonia and meningitis . . . Persons with AIDS often are afraid of staying in the shelters for fear of violence and of exposure to infections such as tuberculosis . . .
(The Medical Director of the Homeless Health Care Project in Washington D.C. cited by R. Mickleburgh, *Globe and Mail* (Toronto) March 25, 1991: A4).

They have difficulty in securing good medical treatment unless they live in large cities. Dr. Philip Berger, an AIDS specialist in Toronto, recognizes that "there should be adequate treatment . . . [in other cities] but the reality is that there isn't." Berger found that the health system is generally not sufficiently flexible to accommodate HIV/AIDS patients. These people are vulnerable to homelessness because they must move in order to obtain adequate care and they lack sufficient funds to cover their medical and living costs
(R. Mickleburgh, *Globe and Mail* (Toronto) March 25, 1991: A4).

While several organizations have been mobilized to support AIDS sufferers, public education remains a significant obstacle. "With almost every individual who has AIDS," according to a specialist at the Toronto Hospital for Sick Children, "there is a horror story about being shunned by friends, families and lovers, about losing jobs, about being evicted." AIDS awareness is high, but drug addiction often guides behavior; 57 per cent continue to share needles and one in three engages in unsafe sex. HIV testing in order to change behavior and minimize the spread of infection often has the opposite effect on homeless people and substance abusers. A medical practitioner at an AIDS treatment center indicated that "the most common response has been to embark on a drug run, get high and escape . . . AIDS completes a triad that destroys all vestige of hope."

Homeless AIDS patients encounter logistical problems in dealing with their illness: they must leave shelters every morning, they must spend hours waiting in line for meals, showers, and bed tickets, there is a scarcity of public bathrooms for people on the street, many of whom suffer from diarrhea, malnourishment, and chronic fatigue, and they have difficulty in taking medication regularly (AZT medication for AIDS patients must be taken every 4 hours). There is often a delay between testing HIV positive and the beginning of primary care, because of a lack of treatment beds. Housing assistance (which is essential to treatment and reduction of stress) often is provided too late.

Once diagnosed as HIV-positive many also have housing problems. A specialist housing association for HIV-positive people in England found that about 10 per cent of its prospective tenants ask to be rehoused because of violence and abuse from neighbors and landlords (C. Partridge, *The Observer*, December 13, 1992: 10). Certain housing associations and voluntary groups construct or acquire units for individuals with AIDS. Some build houses with reinforced glass and steel-framed doors to provide security against attack and to reduce stress (stress speeds the process of AIDS).

In 1993 the National Research Council, part of the National Academy of Science, concluded that AIDS will have little impact on the lives of Americans or the way society functions (1993: 322). Because AIDS is concentrated in "marginalized . . . and socially invisible groups . . . with little economic, political and social power" it is likely that the disease "will disappear" from the public consciousness.

Alcohol abuse Several of the health problems cited above are aggravated by chronic alcohol abuse or drug dependency. It is estimated that about 30–40 per cent of homeless adults are chronic substance abusers, primarily of alcohol, as compared with 10 per cent in the population at large. The figure of 33 per cent, though acknowledged to be less than precise, has been reported in a number of surveys of alcoholism among homeless people (Rossi 1989: 156). This figure, however, obscures gender differences: the prevalence of alcohol abuse is about three times higher among men than women.

Alcohol and health

Alcohol is directly related to serious health problems. The most common causes of death among alcoholics are suicide, murder, accidents, and illnesses including acute hepatitis, cirrhosis, pancreatitis, subdural hematoma, pneumonia, and alcohol-related heart disease.

The problem in ascribing an illness to heavy drinking is that heavy drinkers differ from non-heavy drinkers in other ways. They smoke more. They often eat less. They often lead irregular lives – staying up all hours, never exercising, sleeping it off on park benches. How can these potentially harmful influences be separated from the effects of alcohol?

(Goodwin 1994: 46)

Does alcoholism cause homelessness? It does not for those who are relatively well-off. The poorer an individual is, the more likely problems with alcohol will precipitate homelessness. Substance abuse aggravates other problems. "It is a precondition for vulnerability; it increases the probability that non-economic factors also lead to homelessness" (Conrad *et al.* 1993: 36). It is widely acknowledged that the severity of alcoholism among homeless adults is a consequence of their destitute situations. It is less severe among the newly homeless than among those who have been on the streets for some time. Some individuals "may cope with the privation of homelessness and their damaged self-esteem by drinking yet more heavily" (Koegel and Burnam 1988: 21).

While it is unwise to focus solely on the problems of individuals, and to ignore the systemic problems responsible for recent increases in homelessness, it is impossible to ignore the fact that alcoholism affects many homeless people. Therefore, solutions should include effective and sensitive alcohol rehabilitation programs (Koegel and Burnam 1988: 18). Nevertheless, it is likely that alcoholism will remain common among homeless adults, despite the sums spent on detoxification, which typically does not deal effectively with causal factors. People on the street drink to cope with cold weather, depression, isolation, and physical or emotional pain. Because it dulls pain, induces euphoria, and fills idle time, alcohol is accepted as the drug of choice and as a means of fostering sociability among homeless men and some homeless women. There is, however, no doubt that alcohol contributes to or aggravates health problems, hastens the death of some homeless individuals, and is a factor in accidents and traffic fatalities (when drunken street people are hit by vehicles, for example). People who are inebriated create enormous difficulties for police, care-givers, medical practitioners, and others. This issue represents a continuing dilemma for the operators of shelters, soup kitchens, and group homes as they try to deal with people who are intoxicated, disruptive, and represent a threat to other shelter residents, especially women with children and other adults who are unable to defend themselves. It is relatively easy to bar entry to shelters or to require that all residents of a group home, for example, are "clean and sober," but it may be considerably more difficult to refuse access to someone in the dead of winter when there are no alternative accommodations available and they are likely to end up "sleeping rough" and, perhaps, freezing to death.

The effects of alcohol

Before drugs became a major problem, most police in the United States spent half their duty time dealing with alcohol-related offenses. More than one in five psychiatric admissions are related to alcohol problems. Alcoholics have a death rate at least twice as high as non-alcoholics. One in four suicides in the United States is an alcoholic. In Britain alcohol-induced illness costs 8–15 million days lost from work each year. A British study found that alcohol intoxication was involved in 60 per cent of suicide attempts, 54 per cent of fire fatalities, 50 per cent of homicides, 42 per cent of patients with serious injuries, 30 per cent of drownings, and was involved in 10 per cent of the deaths of people under age 15. Alcohol was associated with 78 per cent of assaults and 80 per cent of breaches of the peace.

(Goodwin 1994: 51–52)

GENDER DIMENSIONS OF HEALTH AND HOMELESSNESS

Health problems are more pronounced among homeless men than women. Men are more likely to sleep rough and are subject to cold injury. Shelters for men are larger and often more crowded; they are more likely, therefore, to contract respiratory and other infectious diseases. Homeless women, however, suffer from an astounding rate of sexual assault, violence, and battering.

Some of the "order of magnitude" observations cited in this chapter obscure marked differences between men and women, especially with respect to alcohol abuse. Of the adults treated at Health Care for the Homeless clinics, 47 per cent of the men but "only" 16 per cent of the women were found to have alcohol problems. Homeless people who abuse alcohol typically are male, most of them over the average age (among homeless individuals) of 35, with high

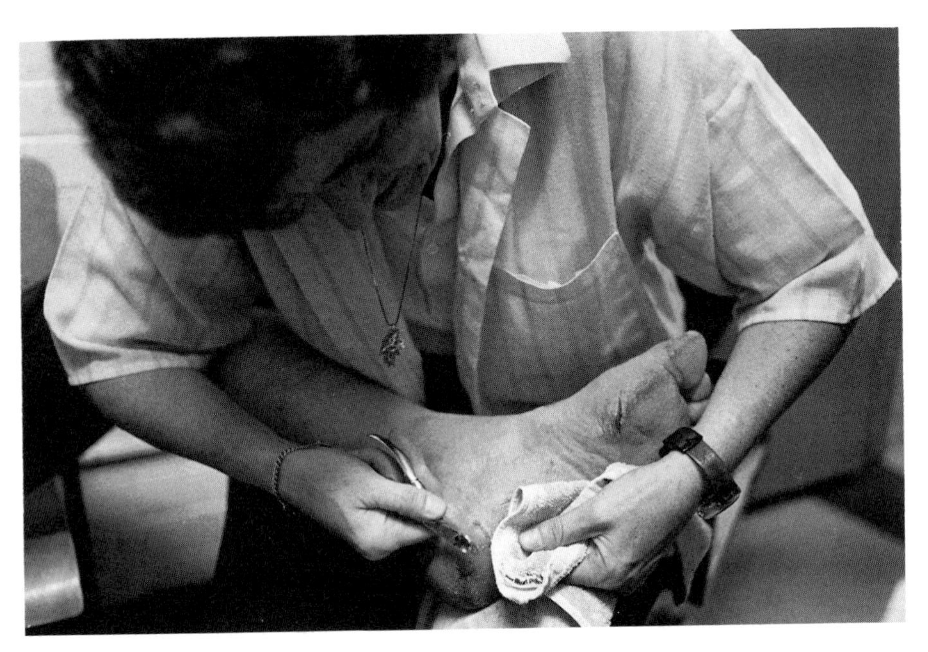

33 Health clinics for homeless people offer first-aid services, referrals and transportation to hospitals, counseling, and assistance in dealing with health and welfare bureaucracies. Clinic workers sometimes assist individuals in obtaining health insurance or social insurance cards, welfare benefits, and documentation in order to qualify for pensions or unemployment insurance (Photograph by Byron J. Bignell 1988)

rates of poor physical, mental, and social health. They have a much higher incidence (than non-drinkers) of neurological disorders, gastrointestinal illness, hepatic diseases, and trauma (Wright and Weber 1987: 68–70).

Women are more likely to be physiologically intolerant of alcohol. Women alcoholics are more likely to suffer from depression and more often have been victims of abuse and have had disrupted early lives. In most cases they are introduced to alcohol later in life than are men and they also become alcoholic at a later age than do men (Goodwin 1994: 64–65).

Marked differences between men and women also are evident in drug abuse. In the United States the incidence of drug abuse is highest among young African-American males, followed by Latino males, then by African-American women, and by white men. The preferred drug varies widely: approximately 37 per cent are dependent on heroin, methadone, or crack, 22 per cent use cocaine, 21 per cent marijuana, and the remaining 20 per cent use miscellaneous drugs (Wright and Weber 1987: 74–76). This pattern changes over time, depending on cost and availability. In recent years the use of crack has increased exponentially. It is made by baking cocaine into crystals with ammonia and baking powder; then it is smoked through stems or glass tubes. It produces a faster high. But it is also the most addictive and deadliest form of cocaine.

Crack addictions

Crack apparently affects women more severely than men. Until its arrival thousands of single mothers who had avoided heroin and cocaine, or kicked the habit, managed to keep their families intact after their husbands or boyfriends had disappeared. Now many of these women live for crack. They resort to prostitution in order to generate cash for their habit. These casual prostitutes – some as young as age 12 – have been mugged, beaten, robbed, and raped while trying to earn money for drugs. Living in "crack dens," they become addicted very quickly and are interested only in acquiring the means to buy another high. Their children are ignored or have been given over to welfare authorities; in some cases, the children have become caregivers in the sense that they are responsible for holding the family together, because parents are absent or incapacitated.

In 1995 a crack-addicted woman in Toronto systematically abused her 6-year-old daughter and eventually killed her.

ACCESS AND ELIGIBILITY

Poor people in general, and homeless persons in particular, have difficulty obtaining access to adequate care. As a result, their health suffers. Delivery of health care and accessibility of services are problematic. Services are normally provided in clinics or emergency rooms of large hospitals. Outreach is not yet common. Homeless persons find it difficult, without a fixed address, to gain entrance to the health care system. Because they lack the resources to maintain personal hygiene, and perhaps because they may appear dirty and unattractive, these people often find that doctors and nurses are reluctant to treat them. Emergency room staff are unreceptive, and homeless transients are frequently treated as abusers of the benefits system.

In 1987 the Canadian Health and Welfare Minister acknowledged that "we don't have equity in terms of healthfulness" between rich and poor. A study in St. Louis found that more than 70 per cent of homeless people "had no usual health care provider and that more than half had received no health care in the previous year" (Wright and Weber 1987: 35). Similar concerns about inequality have been expressed in Britain after the Black Report found that poor health and high morbidity rates are significantly correlated with deprivation (Black 1980; Townsend *et al.* 1985: 662; Whitehead 1987).

One indicator of the degree of commitment to universal health care is public expenditures. Government funding represents 92 per cent of all health spending in Britain, 75 per cent in Canada, but only 42.7 per cent in the United States. Private spending on health care, on the other hand, is highest in the United States, where those with adequate means are able to purchase excellent medical coverage; those lacking the means often must do without proper care. Total spending (public and private) on health in 1990 represented 12.1 per cent of gross domestic product in the United States, 9.2 per cent in Canada, and just over 6 per cent in Britain.

Access to health care programs is a stumbling block. Eligibility for benefits is customarily predicated on illness, alcoholism or drug dependency. As a result, individuals tend to adopt the behavior patterns necessary to secure admittance. Moreover, once admitted, they must remain "ill" or they will be forced to move on. There is little incentive to recover. Access and eligibility are also constrained because public officials frequently limit programs and funding to those with "special needs," such as people with severe physical or mental disabilities. In the United States it is difficult for homeless people to obtain Medicaid unless they are "a parent with a dependent child or completely physically disabled" (U.S. Congress 1990: 38).

SUMMARY

The link between homelessness and health is direct and evident. Many people lose their homes after experiencing serious illness or injury which results in hospitalization, income loss, and eviction. Homeless people are susceptible to illness and injury as a result of physical and psychological stress. Over one-third are in poor health and at least one in three is too disabled to work. People who lack adequate shelter, bathing, and laundry facilities find it difficult to maintain personal hygiene. Their illnesses, skin diseases, and nutritional deficiencies are exacerbated by their lifestyle. They also encounter problems in obtaining access to proper health care and, for those who are visible minorities, disabled, or have AIDS, difficulty in finding housing as well. These obstacles, which are acute for people who have been deinstitutionalized, are described in Chapter 6.

6

DEINSTITUTIONALIZATION

The beguiling notion of a caring community

Much Madness is divinest Sense –
To a discerning Eye –
Much Sense – the starkest Madness –
'Tis the Majority
In this, as All, prevail –
Demur – you're straightaway dangerous –
And handled with a Chain –

(Emily Dickinson)

Among the factors contributing to homelessness are deinstitutionalization and the failure to develop plausible alternatives to institutionalizing mentally ill people. Starting in the 1950s, the British government began closing hospitals for the mentally ill on the assumption that patients would be better looked after outside institutions. Similar policies were pursued by governments in Canada and the United States in the early 1960s. The concept of deinstitutionalization was predicated on an optimistic belief in the compassionate capacity of the community, on the discovery of psychotropic drugs, and on the belief that large medical and psychiatric hospitals failed to provide tangible benefits to most patients.

MENTAL ILLNESS, DEINSTITUTIONALIZATION, AND HOMELESSNESS

It is difficult to estimate the extent of mental illness among homeless people. Part of the problem is differentiating between substance abusers and others. The link between homelessness and mental illness is the subject of numerous research and epidemiological studies – but the results are mixed (Blankertz and Cnaan 1994: 537–538). Bean concluded that the overlap between homelessness and mental illness remains "a tangled web" of confusion (Bean *et al.* 1987: 411–416). Mental illness may either cause or result from homelessness. Several of the characteristics of the mentally ill are common among homeless people: they are poor, marginalized, often in poor health, lacking social supports, and many are members of minority groups. The picture, then, is muddled. It is now generally accepted, however, that about one-third of homeless people require psychiatric help. Many of these individuals became homeless after being deinstitutionalized. Others would likely be institutionalized if adequate facilities and funding were available (Fischer and Breakey 1991: 1121–1124).

Some critics suggest that the common image of homeless persons as mentally disturbed is exaggerated, representing an attempt to "medicalize" the issue. Advocates believe that the tendency to focus on individual flaws, implicit in this medical model, restricts attention to those who are mentally or physically ill. While stereotyping these homeless individuals as "the deserving poor," this approach allows government to absolve itself by focusing on personal illnesses rather than on underlying social and economic problems.

Schizophrenia

Rates of schizophrenia do not change, but the rate of hospitalization for schizophrenia and other psychoses has changed. In 1955, there were 93,000 patients in New York State asylums; last year there were 11,000. Where have the remaining 82,000 and their descendants gone? In one generation, a flood of pathetically ill people has washed onto the streets of America's cities. We step over these wretched and abandoned folk sleeping in doorways and freezing on grates. They, too, have become accepted as part of the natural landscape.

(Krauthammer 1993: 20)

Those who have been deinstitutionalized are irrevocably bound to society in a dependent relationship based on illness and the dispensation of charity. Once these individuals return to the community most are not capable – initially at least – of functioning on their own. One person who was in a psychiatric hospital seven times recalls

> the trauma [which arose] from separation anxiety . . . I remember being unable to cook or shop or participate fully in life after discharge. There was a perceived need to carry around the walls of the institution because it was safer that way.
>
> (Canadian Council on Social Development 1985)

After extended periods of enforced dependency, they face a frightening maze of bureaucratic hurdles. A great deal of time – if they are physically able and mentally alert – is spent in making the rounds of agencies. If an individual lacks a birth certificate it is difficult to obtain health care, welfare payments, or other benefits. In Canada it is necessary to pay an application fee in order to obtain a birth certificate, which is a prerequisite for securing a health insurance number or a social insurance card. Both are needed to get a regular job.

For those who are disoriented, have lost touch with their families, have misplaced their papers, or had them stolen – a common occurrence in emergency shelters – life becomes a quest for documents and credibility. Immediately after being discharged they are usually unable to deal with this process without assistance. Their feelings of inadequacy and frustration are reinforced by this system. Once caught in the revolving door of institutionalization-deinstitutionalization, escape is difficult.

THE RESULTS

Since 1952 more than 100,000 long-stay patients have been discharged from asylums or psychiatric hospitals in Britain. Life expectancy is lower among these released patients than those in hospitals (Haughland *et al.* 1983) and many patients have not survived. A survey in Britain found that

> a fifth were not receiving benefits, half were not receiving medical or psychiatric care and a third obviously required treatment for physical problems, including tuberculosis, which affected about one in five of the 25,000 men who used the Camberwell Reception Center in the last two years prior to its closure . . . Despite undoubted advances in our understanding of psychiatric illness, there are obvious defects in our system of care.
>
> (Weller 1991: 45)

The savings realized by the closure of institutions for the mentally ill were not redirected to special-needs housing or to community care: the National Federation of Housing Associations estimates that only 9 per cent (a total of 4,050 beds) of special-needs housing schemes in England are allocated for people with mental illness.

The British government plans to close all psychiatric hospitals by the end of the century, yet has plans for twenty-six new prisons. Many former patients are now in bed and breakfast hotels or "homes" which exist primarily to reap the benefits of monthly payments from the public treasury. In Portsmouth, for example, there were more than 900 such establishments at the end of the 1980s. The number had multiplied five times during the decade as landlords realized the attraction of a clientele which offered the prospect of regular rent, year round.

In the United States, even chronically mentally ill people have been deinstitutionalized. After a period of freedom they usually end up in hospital or jail, rather than being readmitted to a mental institution. Many former psychiatric patients wander about aimlessly (Belcher and Toomey 1988: 145–153). A Los Angeles survey found that people who had been hospitalized with psychiatric problems

> were the least likely to sleep in an emergency shelter, had been homeless nearly twice as long as the rest of the sample, had the worst mental health status, used alcohol and drugs the most, and were the most involved in criminal activities.
>
> (Gelberg *et al.* 1988: 191–196)

In some cases, deinstitutionalized patients are prematurely released, without proper investigation, to the custody of their families or to foster family members; but many of these people are not prepared to take on the responsibility of caring for schizophrenic children or spouses. The belief in a caring community, evoking images of neighborliness and social supports in a village-like setting, is ill-founded.

> It has served to mask conditions in private custodial settings such as rest homes and boarding homes which have emerged as . . . the largest type

of accommodation which caters to those single individuals who are non-competitive in the labor force and require some level of on-going support.

(Association of Municipalities of Ontario 1986: 1)

Nevertheless, deinstitutionalization continues. In Ontario the total number of patients in mental hospitals declined by 75 per cent between 1965 and 1976 (Gerstein 1984: 25). In the United States the population of state mental hospitals dropped by more than 80 per cent after 1955. In New York State about 8,000 people a year, who would have been admitted under the old policies, no longer qualify for institutional care. This marginalized population continues to grow. Many of those discharged or refused admission have no place to go (U.S. Congress 1983: 903).

The "revolving door" of homelessness

Marcie is age 33, single, a product of a broken home, who worked regularly at good rates of pay until hospitalized after an automobile accident. As a result of the accident she became addicted to pain-killing drugs. Since her hospitalization she has been in and out of courts, hospitals and detox centers, and has become suicidal as well as violent. The following timetable follows her on her trips through the revolving doors of these institutions over the past seven months:

January: Marcie is hospitalized overnight for a drug overdose. Later in the month she is caught shop-lifting, is scheduled for court, but fails to appear and is arrested.

February: While at the emergency shelter she appears to fall into a coma and is rushed to hospital, but is discharged later the same evening.

March: Admitted to detox center, she is later compelled by a court order to enter a mental institution, but is released in less than a week. Becomes violent and beats up her landlord.

April: Evicted from her apartment after fight with boyfriend and setting fire to her mattress. Barred from women's hostel because she overdosed on drugs.

May: Attempted suicide. Admitted to hospital. Released on second day.

June: Showed up at night shelter in battered condition. Taken to emergency ward of hospital. Suffering from depression.

July: Violent argument with staff workers at women's shelter. Barred from the shelter for one month. Later has similar problems at emergency hostel.

Government agencies play a key role in this process. Deinstitutionalization, government policies of fiscal restraint, and judicial decisions restricting agencies' ability to institutionalize individuals, have made admittance to psychiatric hospitals problematic. Beginning in 1963, the Veterans' Administration (VA) reduced its inventory of psychiatric beds by well over half. The VA has been criticized by Congress for its passive approach to care for veterans. This hands-off policy is particularly inappropriate for poor, uneducated, and mentally ill veterans who neither know their rights nor how to manipulate the system. Veterans without a high school diploma are 25 per cent less likely to have contact with the VA than are veterans with a college degree. Two studies conducted in New York, one by the State Office of Mental Health, found that only about one in eight homeless veterans receives benefits of any sort from the VA (U.S. Congress 1983: 907–914).

Where do these people go after being discharged or refused entry to hospitals? Many in the United States find their way to hostels, emergency shelters, or rooming houses, if space is available. Some will end up in welfare hotels or motels; others are jailed or imprisoned. In Britain, extensive use is made of bed and breakfast hotels, group homes, hostels, and single room "bed-sits." In 1969 there were 14.6 places per 100,000 adults in England in staffed group homes and hostels for people with psychiatric problems. By the early 1990s this figure had tripled to approximately 45 places. These unsupervised facilities tend to become institutions themselves, with fixed meal times and bedtimes and standardized procedures. In large cities, ex-patients are routinely housed in empty council housing, including partially abandoned tower blocks awaiting demolition or rehabilitation. Life in these high rises is far from ideal. Vandalism, drug trafficking, and muggings are commonplace.

The reaction of some local governments to homelessness is to incarcerate vagrants or to escort them out of town. In New York City the municipal government concluded that the homeless mentally ill were slowly killing themselves on the street. As a result, they initiated Project Help; homeless people were forcibly removed from the pavement, assessed, and confined for treatment without their consent. This program raised serious ethical questions. The American Civil Liberties Union went to court to secure the release of one homeless woman, Joyce Brown, who subsequently gained widespread exposure and joined the lecture and talk show circuit. A few months later, after being picked up, unconscious and suffering from epileptic seizures, she was again institutionalized.

In Britain the 1983 Mental Health Act protects the rights of individuals and does not permit forcible confinement. Hospitals are reluctant to take on long-term patients; they continue to discharge mentally ill people.

In Canada the prospects of finding suitable accommodation after being released from a psychiatric hospital are limited. According to Dr. Samuel Malcolmson, Chief Psychiatrist at Toronto's Queen Street Mental Health Centre, those with serious mental health problems should be looked after only in hospitals or in supervised group homes. But most group homes refuse to accept individuals prone to substance abuse or violent behavior. As a result, mental health hospitals now tend to keep patients longer. The opportunity costs of this pattern are formidable: major clinics like the Queen Street Centre in Toronto turn away half of their applicants.

Frequently, discharged psychiatric patients end up in boarding houses which are unregulated regarding care standards and staff-to-patient ratios. In most jurisdictions these places are not covered by landlord and tenant codes and residents have no security of tenure. Troublesome residents are routinely sedated and confined to their rooms or strapped to wheelchairs.

This situation prompted reforms in Ontario, which passed the Residents' Rights Act in 1994. It extends the coverage of the Landlord and Tenant Act to people in care homes, transitional housing, and non-profit projects. They now have security of tenure, the right not to be evicted except on specific grounds and through due process, and they have all the accompanying rights to privacy, proper maintenance, and peaceful enjoyment of their premises. Coverage of the Rent Control Act also was broadened to include the housing portion of the monthly payments made by these residents; however, this does not cover non-housing services such as meals, medical care, or security, which may be increased after a 90-day notice period.

ALTERNATIVES TO INSTITUTIONALIZATION

Britain, the United States, and Canada pursued deinstitutionalization policies without widespread success because inadequate attention has been given to community care alternatives which meet the needs of those who are discharged from institutions. These requirements include decent housing, accessible health care and community services, and educational, vocational, and recreation programs. A flexible community services model which encourages self-help and builds self-esteem within a supportive setting satisfies these objectives.

Alternatives include the "clients movement" which challenges the seclusion of patients, excessive use of medication, and the hierarchical nature of the client's relationship with the medical profession. This is referred to as a "valorization of the subject" which highlights skills and potential contributions rather than concentrating on disabilities (Mezzina *et al.* 1992: 69). The movement is predicated on the use of consumers' own social skills as a means of rehabilitation, focusing on people's strengths rather than on weaknesses or deficits that must be treated and overcome. It seeks to overturn the notion of mental illness as an "I am illness, one that is joined with social identity and perhaps with inner self, in language and in terms of reference" (Sullivan 1992: 205).

One form this movement takes is the promotion of self-advocacy and self-help. In Kansas, initiatives focusing on the strengths perspective have resulted in recidivism of 15.5 per cent, half the rate of traditional deinstitutionalization approaches. One of the keys is to concentrate on social rehabilitation and reintegration. Without this focus proliferation of services may be counterproductive:

> comprehensiveness may very well lead to a new much more refined "total institution," encompassing all problems of all people in all kinds of programmes . . . without ever bringing the clients back into the community or stimulating them to take care of their own lives.
>
> (Schnabel 1992: 65)

In some programs consumers are being successfully employed as service providers within a team working with homeless people who are severely mentally ill; a related approach is being pursued by consumer self-help organizations (Dixon *et al.* 1994: 615).

Another key element is work, which "has instrumental and symbolic importance and signifies that the individual is a fully participating member of society" (Sullivan 1992: 207). Meaningful employment may improve self-esteem, quality of life, and often helps former patients to increase their social involvement and management of their illness and their lives; lack of these elements, conversely, contributes to chronicity. A study in New Hampshire with a supported employment program found that 28.3 per cent of previously unemployed individuals obtained competitive jobs in the community after the program, compared with only 8.2 per cent of those who had been in treatment (Drake *et al.* 1994: 526). There is now substantial experience with alternatives to institutional care. A number of North American studies reveal that hospitalization became self-perpetuating and was more costly than community care alternatives in both economic and social terms. Quebec researchers found that the cost of institutionalizing a person was about twice as much as helping them stay in an apartment where extensive support services were available.

A widely cited study in the United States found that an experimental community-based approach to mental health treatment cost about $800 more per patient in the first year than a traditional hospital-based approach. But the alternative program resulted in productivity gains of $500 per person. Moreover, patients cared for in the community showed improvement in planning and decision-making, their symptoms of mental illness diminished, and their satisfaction with life increased (Weisbrod *et al.* 1989: 405).

SUMMARY

In the past three decades governments in Britain, Canada, and the United States decided to discontinue asylums and psychiatric institutions or to severely reduce their size. In many instances, this became a budget reduction exercise. The money saved was not redirected to community services nor was it made available for housing to accommodate those released from institutions.

Deinstitutionalization is predicated on misleading notions: that psychotropic drugs could stabilize patients to the extent that they immediately would become self-sufficient; and that the vague concept of community care was preferable to institutional incarceration. It is apparent that deinstitutionalization, as part of a broad-based effort to reduce the size and scope of the welfare state, may become a guise for privatization. Patients released (or potential patients who are refused admittance) are cast adrift without adequate supports. Many who are released find that emergency shelters or hostels are the only available accommodation.

Despite this poor record, it is clear that well-supported community care programs are preferable to institutionalization. Individuals generally fare better in community-based programs, become more self-sufficient, and display fewer symptoms of mental illness. Moreover the long-term cost of community-based initiatives may be less than institutionalization.

7

THE EXPERIENCE OF HOMELESS PEOPLE

> I feel anger toward an uncaring society . . . to be homeless means that you have no rights, but homes are a right, not a privilege.
>
> (Woman at a London shelter)

> Several weeks ago, I was too late to get a bed in any of the downtown hostels. I made a tour of the all-night coffee shops. Walking around, I noticed that the bus shelters I passed on Queen St. were occupied, as were many doorways. Around 3 a.m., I ran out of coffee money, so I headed for a quiet park to lie down. Of course, it started to rain. Well, in the course of the next hour, I scrambled alleyways, fire escapes, across warehouse roofs – and without exception, every little hiding hole was filled with some unfortunate person like myself. Finally, I saw some piles of cardboard by some tractor-trailers. I pulled the first pile aside. Someone was sleeping underneath. The same thing with the second pile. And the third had two people under it. All this within fifty feet of trendy Queen West.
>
> (*Rumours*, Toronto, 1986)

LIFE ON THE STREETS

For people without housing who live on the streets, their days are marked by endless walking and waiting. They keep moving, often at the request of police, shelter operators, or store owners. They stand in lines waiting . . . for shelter, food, welfare benefits, and health care.

> I need a place to stay. My legs are badly ulcerated . . . Not the shelters or the flops, they are not for me. The police took me off the subway and to the Men's Shelter a few months back, but I couldn't stay. I am 68 years old and I can't defend myself down there.
>
> (Coalition for the Homeless 1983: 17)

Tenants sometimes become homeless after being intimidated by landlords who evict them in order to convert their buildings to condominiums. Landlords may threaten legal action, fail to make repairs, shut off heat and water, illegally lock residents out, and arrange burglaries, arson, or physical assault to convince sitting tenants to leave. Homeless individuals are victims of harassment and violence because they are vulnerable, often alone, and lack political clout or even civil rights. Much of this crime is unreported because

Women who use shelters in American cities

- Over 90 per cent of the families in shelters are headed by women; their median age is 27. Women represent one-third of homeless under age 29.
- Three-quarters are non-whites; over half are African-Americans.
- More than two-thirds are living in poverty and are dependent on public assistance; one in four is employed (Breakey et al. 1989: 1352).
- They moved an average of four times in the past year; over half lived in shelters or welfare hotels during the previous five years (Bassuk and Gallagher 1990: 20).
- Most are not currently married; two-thirds have children; the average is two children per family (Bassuk and Rosenberg 1988: 783).
- Their support networks are limited. Only half have at least monthly contact with other family members; by adolescence, two-thirds of the mothers were living in female-headed households.
- Most have less than a high school education.
- One-third are currently abusing drugs or alcohol. Four out of ten single women have been treated for drug abuse; but only 11 per cent of women with children use drugs or alcohol frequently.
- More than two-thirds are depressed. Over one-quarter have previously been hospitalized with mental illness.
- One-third are chronically homeless, one-third intermittently homeless, and one-third newly homeless (Mulburn and Booth 1989–1990).
- Shelters for women often exist on donations and cannot provide child care or fulfil the nutritional needs of pregnant women or children. Mothers are sometimes required, as a consequence, to put their children in foster homes (Reyes and Waxman 1989).

of public indifference. Chronically homeless individuals, many of them elderly or disabled, are vulnerable to crime because of fatigue, weakness, or malnourishment. Both heterosexual and homosexual rape are common (Kelly 1985: 77–91). Shelter residents also may be the perpetrators of crime, preying on other shelter occupants who are defenseless.

When homeless persons are apprehended by police, it is usually for victimless crimes: three-quarters of the arrests recorded against them in Baltimore were for disorderly conduct, trespassing, or sleeping in public parks. An additional 15 per cent involved offenses against property, some of which were for stealing food (Fischer 1988: 46–51). Homeless people can expect some harassment from members of the public, shop keepers, downtown business owners, and police. Often municipalities impose fines for vagrancy, turn on lawn sprinklers in public parks at night, place barriers across park benches, and take other actions designed to discourage sleeping in public places. Since the mayoral election of 1991, won by former police chief, Frank Jordan, the City of San Francisco has cited over 4,000 homeless people and imposed a $71 fine on each; when they are unable to pay, an arrest warrant is issued. The National Law Center on Homelessness in Washington, D.C. found that thirty-two of forty-nine American cities surveyed in 1994 had enacted laws specifically prohibiting such activities as panhandling and sleeping in public places. In numerous jurisdictions fingerprinting is mandatory for recipients of General Welfare Assistance.

Comments by participants in the Homeless Persons Outreach Project (City of Toronto, 1990)

- *Advocacy* We can speak for ourselves; we don't need people to speak for us. We should have more input.
- *Community Services* They are designed by staff and are operated on a timetable to suit staff people. They don't necessarily match the needs of those who are supposed to be served.
- *Food* In downtown areas, groceries are expensive and alternatives are few. Food served at hostels is inadequate. At the food banks, too many embarrassing or hostile questions are asked, especially of young men.
- *Justice system* Not perceived to be just. High fines are imposed on people who cannot pay. They are then sentenced to jail.
- *Lack of power* They are exploited by cheque-cashing services that extract exorbitant fees. Because they lack proper identification they are unable to open bank accounts. Labor pools "use" them by paying minimum wages and imposing spurious fees. They feel that hospitals ignore them and offer inadequate treatment. Many believe that city planning and decision-making excludes them. Living in rooming houses means high rents, absentee landlords, substandard conditions, too many rules, no visitors permitted. A number complain about abusive police who harass them for sleeping under bridges or in parks. Politicians, bureaucrats and social service providers are often unresponsive to their needs.
- *Major problems* Personal safety on the streets, in hostels and while using (or waiting for) public transit.
  Waiting for community services; exemplified by "Being put on hold for the suicide line."
- *Suggestions* Let community groups fix up old buildings and let people move in . . . Provide a safe, clean environment with accessible parks and open space . . . Include homeless people in planning for transit, social services and the built environment.

Homeless persons who are able to do so seek alternatives to the shelters. Parks are used if possible, as are beaches, woods, ravines, alleys, abandoned vehicles and boxcars, school buses and buildings, as well as storage containers and spaces under bridges or freeways. Some even live in subway or train tunnels (Toth 1993). Others opt for all-night movie theaters, coffee shops, bus terminals, or train stations, while a few spend mind-numbing hours riding public transit.

Single homeless people in particular depend on an institutional network for survival (Laws 1992: 119). They are frequently on the move but remain within a relatively small area of the inner city which is strictly circumscribed (by zoning and by community opposition in adjacent areas) and defined by the location of interdependent institutions (Wolch and Dear 1993). Included are: rooming houses, SROs, emergency shelters, hostels and missions, drop-in centers, food banks and soup kitchens, the Salvation Army or clothing depots, check-cashing places, pawn shops, casual labour pools, and public baths. These areas tend to have a concentration of retail outlets which are heavily used by street people (doughnut and coffee shops, beer and liquor stores, and taverns), public facilities (libraries and public transportation) and support

Profile of a family becoming homeless

Growing up in a dysfunctional home was "NO FUN," to say the least. The scars will never go ... Hitting on the back with a belt, or a fist, or a slap across the face were somehow overlooked. I grew up afraid to achieve anything ... If social service agencies, as we have today, had been around when I was a child maybe someone would have seen the troubles and helped. I see so many people everyday that are just now realizing their scars and are paying the price for it as adults ... just as I have and will. At 28 years of age the real collapse hit me ... the problem with it then was the fact that I had 3 children and a wife. Duty or not, I collapsed ...

When my family and I became homeless the childhood memories of being the man of the family changed dramatically ... I didn't feel like the man of the family anymore ... I kicked myself each day and kept repeating to my wife to just leave me and go to her parents' home ... Glad now she didn't ... The nights were very frightening [sleeping outdoors] ... the most dangerous elements from society are out there ... I didn't sleep good, not one night. Not only had we lost every material possession, gotten dirtier each passing day, but couldn't sleep to escape it for awhile.

Resentment becomes easy ... I was ashamed of myself ... Some of the other people that were also homeless would go buy beer or whatever and, at first I won't partake ... after a while I did ... they didn't look down at me, they just wanted a friend and a drinking buddy. I could at least pretend that I didn't hurt and that somebody liked me even if I didn't like me.

(Terry McClintic, July 28, 1995)

services: hospitals, street health clinics, counseling and detoxification centers, welfare and employment offices. Dependent on walking or on public transit, their food shopping choices are limited. Because supermarkets in North America have fled to suburban locations, most inner city residents must rely on the more expensive "convenience stores."

Women and men differ significantly in the nature of their networks, activity patterns, and the ways in which they use the city. Men spend more time on the streets; 89 per cent of men are alone on the streets, while 55 per cent of women are accompanied (89 per cent of them by their children). Shelters for men, almost without exception, require that they leave early each day and they are not allowed to return until late afternoon. Women's refuges allow them to remain within the sheltered surroundings and permit long-term stays; they also provide a number of services under one roof, while traditional hostels for men offer only space for sleeping and, perhaps, eating and bathing. Men range more freely throughout "their" part of the city, making extensive use of labor pools, parks, and other places where it is possible to congregate or to sleep undisturbed, at least for short periods. Women are more restricted in their movements, because they lack the safety of numbers that some males enjoy, and because they are more vulnerable and must be more concerned about their physical security.

People on the streets supplement their incomes by piece work, distributing pamphlets, and, in some cases, by prostitution. Most homeless people do not panhandle regularly. In fact, most panhandlers are not homeless. Before the onset of AIDS, some regularly sold their blood and many still make donations (for a fee) to the plasma center twice a week (Snow and Anderson 1993: 154).

Others participate in casual labor arrangements for jobs which last only a few hours. Those seeking work must queue up in the early morning in order to be selected. Frequently the work is dirty or dangerous – smashing automobile batteries, for example – and protective clothing is not provided. If injured they have no recourse because they are not registered for disability or worker's compensation. Some employers renege on oral agreements and refuse to pay the full amount owed for work completed. Often they withhold part of the pay for transportation or other charges. Remedies are possible, however; in 1995 Florida passed a law, drafted by the Legal Services Homeless Task Force, which makes it illegal for labor pools to charge workers for safety equipment, for cashing paychecks, or for more than the actual cost of transportation or work clothing.

THE FEMINIZATION OF POVERTY

"Housing is a feminist issue" (Miller 1990: 2). The feminization of poverty is common to Britain, the United States, and Canada. Women in poverty are confronted with the prospect of alienation and homelessness. Homeless women violate social norms: they are alone, apparently not dependent on any man, oftentimes dirty. They obviously do not conform. Public reactions range from bewilderment, disgust, rejection, to ostracism. Harris noted that these women are cut off from the larger society, from their personal support networks, but also "are often fundamentally alienated from their own core selves" (Harris 1991: 16).

In the United States about 30 per cent of all households are headed by women; their income is less than half of the national average. More than half of all families in poverty are headed by women. A single mother in a minimum-wage job earns 20 per cent less than what is defined as a poverty-level income for a family of three; a mother on welfare receives benefits which average about 55 per cent of the poverty level. Over half of the nation's low-income single mothers have housing problems in terms of crowding, affordability, and/or physical inadequacy.[1] Female home-owners, many of them elderly, are twice as likely as the general home-owning population to have affordability problems.

In Canada approximately one in three marriages ends in divorce. This trend spawned a rapid increase in the number of single-parent families during the 1980s; over 85 per cent of these households are headed by women and more than half live in poverty. Women represent the great majority of the frail elderly group. Most older women living alone are on low fixed incomes. Women face discrimination in the marketplace due to low incomes, lack of a regular work or credit history, and a shortage of affordable housing. Among their concerns are safety, security, access to public transportation, and day care. About 65 per cent of Canadian women are renters, compared to 29 per cent for men. Single women spend 50 per cent of their income on housing; single mothers spend 65 per cent. Women comprise 60 per cent of the residents in publicly assisted housing and are the principal recipients of public income supports. When cutbacks are made, jeopardizing support services and assisted housing, women on the margins become more vulnerable to homelessness (McClain and Doyle 1983).

The 1991 Census of England and Wales (OPCS 1991: 23, Table 4) found that there are almost 19 million households in England and Wales; 30 per cent are headed by women. While three-quarters of men are home-owners, less than one-quarter of women own their homes. Many of these are older women who have outlived their spouses; because they are home-owners they generally do not qualify for state assistance. Yet they are often in older homes with maintenance problems that are beyond their capabilities; and many have debt problems (Gilroy 1994). Women are vulnerable to homelessness because they are poor; according to the Equal Opportunities Commission (1993: 11) they earn about two-thirds of average male earnings, while still maintaining primary responsibility for child care and for dependent relatives. The 1991 Labour Force Survey (quoted in Gilroy 1994: 46–47) indicates that 42 per cent of women in waged work are part-timers, compared to only 4 per cent of employed men. The London Housing Unit (1993) calculates that only 17 per cent of women working full time earn enough to purchase a three-bedroom property; slightly less than one in three has sufficient income to purchase a two-bedroom; almost half of all women working full time do not earn enough to purchase even a one-bedroom property, leading one observer to conclude that "the woman has to find her prince before she gets her palace" (London Housing Unit 1993 quoted in Gilroy 1994: 54). In any event, home ownership is not a panacea and may not be an appropriate solution for low-income households who would have insufficient funds left over for essential purchases; they may be unable to maintain mortgage payments, particularly when only variable rate mortgages are available.

Despite the dramatic rise of housing needs of non-nuclear households in Britain – nuclear families now represent less than 30 per cent of households – public housing policy does not favor the provision of housing for single women or non-traditional families. Sheila McKechnie, former Director of Shelter, concluded that:

> Housing is about social power. When access to housing becomes based almost entirely on income and not on need it is not hard to imagine who loses out. It is the same group that always loses out when individual wealth replaces collective responsibility . . . women. Not all women, of course, but some women more than others. But they lose out because of the fundamental inequality of being a woman.
>
> (Miller 1990: 175)

Men and women are homeless for different reasons; their housing needs are different; and they require different approaches and supportive services to address those needs (Leavitt and Saegert 1990; Harris 1991; Golden 1992; North and Smith 1993). Men usually succumb to literal homelessness because of loss of employment, leading to loss of housing and, perhaps, family connections, all of which may be aggravated by substance abuse or some form of illness. Homeless women are often referred to as "situationally homeless"; that is, they are on the streets because of immediate economic problems, because of mental health problems, or because they have been abused. Once on the streets, of course, it is highly likely that physical and sexual abuse will continue, and perhaps worsen, unless they can gain entry to a women's shelter (Ambrosio and Baker 1992: 51). Women are younger, and most are non-white. Despite being jobless for long periods, women (especially those accompanied by children) are homeless for much shorter periods than are men (forty-one months for men, thirty-three months for single women, sixteen months for mothers with children). They are less likely to have institutional contacts with mental health or criminal justice systems (20 per cent of mothers with children, 50 per cent of single women, 74 per cent of single men) (Baker 1994: 478). Women as a group are in greater need than men of most social services. Single women need more intensive psychosocial counselling and alcohol treatment than do women with children. The latter need subsidized housing, social benefits, parenting skills training, child care, and job training (DiBlasio and Belcher 1995: 131–137).

Given the feminization of poverty, why are women less likely than men to become homeless? This is a complex question which points to the social construction of homelessness. Part of the answer lies in "snapshot" or point prevalence surveys which are more likely to capture the chronically homeless (mostly men) and to concentrate on people living on the streets (virtually all are men) and on large shelters, many of which are for men only. Men have a higher incidence of homelessness, in part because the loss of SROs and other cheap accommodations, resulting from gentrification or renewal, affects mostly low-income men; they are left with few alternatives other than to resort to shelters or the streets. Women sometimes are not counted as homeless because their refuges or transitional homes are often excluded from surveys; residents of these places are not considered to be absolutely homeless. Moreover, women are recognized as a priority for housing in most cities, and are more likely to qualify for shelter subsidies and to be housed by municipalities (Burt 1992; Liebow 1993: 237). Women have a much higher probability of qualifying for welfare, because they are usually accompanied by children; this assistance is normally used to secure housing. Women are more likely to be attached, either to children or to other family members or to a support network, even if that social network is confined to Skid Row or to an informal encampment (Rowe 1988: 36–38). They have a better chance of receiving assistance from kin than do men, in part because men have a higher rate of substance abuse and personal disabilities, which cause them to be rebuffed; and, as well, because kin-based norms of obligation are more narrowly construed for men than for women. Finally, women are less likely to describe themselves as homeless; they do not, as a result, foreclose their options and thus are unlikely to succumb to literal homelessness because that condition places them at great physical risk (Watson and Austerberry 1986; Baker 1994: 482).

Recent studies in the United States found that more than 40 per cent of homeless women reported abuse from a spouse, and 39 per cent of that group had experienced more than one abusive relationship (North and Smith 1993: 427). A substantial percentage are turned away from shelters because of a lack of beds, and one in three abused women return to the abuser because they cannot locate housing (Bard 1994: 13) – hence, the growing need for places that cater specifically for women fleeing an abusive partner. Typically, they require refuges with supportive services that permit long stays and allow them to keep their children with them. There is evidence that these women respond well to small, family-like hostels which foster a sense of community and allow all residents to participate in decision-making. In light of this evidence it is difficult to understand why some housing providers continue to place single women and single men in the same projects, when all they have in common is poverty and lack of shelter (Brown 1995: 92).

Abused women

Malika Saada Saar, coordinator of the Family Rights and Dignity Program for African-American single mothers in San Francisco, is convinced that

> You cannot separate our invisibility from our race and our gender. In our society, women are told that if you are silent, you are strong; that what gives us strength is that we shoulder our community's problems and we do it quietly. Our silence is what forces us to stay with crack and to stay with an abusive relationship. For women – and especially for women of color – to raise their voice as a community is a revolutionary act.
>
> (Mesler 1995: 5)

Women who are abused usually leave home abruptly after being beaten on a number of occasions. As a result, they do not anticipate being homeless. Most see their situation as being temporary and are unable to directly address their problems. Denial is a common coping mechanism. Many are paralysed by fear. Wherever they go, they feel unsafe, afraid that their abuser will find them. As a result, most shelters for women do not reveal their locations and will not admit male visitors. Many battered women feel safe only in the shelter, surrounded by people who will protect them. Recently, certain jurisdictions have made arrest mandatory when police suspect that an assault has occurred. This gives women more legal protection. Ironically, though, it may cause increases in the numbers of homeless women; demand is growing for shelters that specifically meet the needs of women (and women and children) fleeing abusive relationships.

Many battered women have been out of the job market for some time, so their earning power is reduced. They also face difficulties in locating suitable housing; landlords commonly discriminate against single mothers with children. Moreover, most women who suddenly find themselves on the street or in a shelter for battered women are devastated; it takes them months to come to grips with this crisis.

THE CHILDREN OF POVERTY

Among those who have suffered most from increasing homelessness are children. About 12.5 million American children live in poverty. One child in eight has no health insurance, only one in five has adequate day care, one in seven will be a school dropout, and one in six lives in a family where nobody has a job. Only 16 per cent of those eligible for Head Start (a federal program which provides schooling and breakfast for poor children) in the mid-1980s were enrolled in the program (Edelman 1987: 29–31).

Studies in Philadelphia and other cities demonstrate that the effects of homelessness on children include poor health, severe disruptions in family life, delays in physical and emotional development, and poor school performance (Fox and Roth 1989: 141). Over 40 per cent must repeat at least one grade, almost half do not attend school regularly, and a great number need special education classes. Abuse is common. Women in shelters report a high incidence of family violence, first from their fathers and later by husbands or boyfriends. Subsequently, many women are violent with their children, and many are under active investigation for child abuse and neglect (Bassuk and Gallagher 1990: 20).

In Britain the Health Visitors Association found these children more vulnerable than others in their age-group to accidents (burns and falls) and to disease. They are prone to vomiting (because of a constant diet of greasy, fast food), chest complaints (from dampness), and scabies. The study discovered that half of the children under age 5 had not been immunized against diphtheria or polio. Their development is retarded because of lack of play facilities or even space to learn to walk.

Street children

The street offers its children the spectacle of society without integration into its values: proximity, but not participation. It becomes symbolic of their distress. It replaces school, and has a very different syllabus. It belongs to everybody and nobody, and puts everyone on the same footing. It cancels out the past and makes the future uncertain: only the present moment counts.

(Agnelli 1986: 9)

In Canada children are often the victims of poverty. Among lone-parent immigrants with young children, for example, 58 per cent are below the low income cut-off set by Statistics Canada.

THE DILEMMAS OF HOMELESS YOUNG PEOPLE

Evidence of a "throwaway youth culture" surfaced in the 1980s. The U.S. Department of Justice (Office of Juvenile Justice and Delinquency Prevention) estimated that the country had 59,200 displaced children, mostly teenagers, in 1989. There are 700,000 young people taken into America's juvenile justice system each year (U.S. Senate 1993: 2). In Toronto more than one in four individuals using shelters are between the ages of 16 and 24. Most are forced to leave home because their families have broken up or because they are the victims of physical, sexual, or emotional abuse. Young people who grow up in foster homes often become transient. After leaving foster care they frequently move from one group home to another. A Canadian survey revealed that 20 per cent had lived in three or more homes within the past year (Municipality of Metropolitan Toronto 1995).

The National Health Care for the Homeless Project in the United States found that homeless girls, ages 13–15, are fourteen times as likely to become pregnant as their non-homeless peers; almost one in three homeless girls between the ages of 16 and 19 get pregnant. Three-quarters of all missing children and youths in Canada are runaways. The Social Planning Council of Winnipeg found that one in three runaways contracted a sexually transmitted disease, 60 per cent put themselves at risk of contracting HIV/AIDS, and one in five exchanged sex for shelter. The average first-time runaway on the streets is a 12-year-old girl, who is a victim of sexual abuse, is using drugs or alcohol, and has attempted suicide at least once. Most run from a home where parents abuse alcohol or drugs. Typically, they are escaping intolerable domestic situations, but they generally are not aware of what awaits them on the street. Runaways resort to prostitution, drug dealing, shoplifting, or other illegal means of support (Bublick 1991: 34). Public officials frequently advocate returning young people to their parents; but many have no real homes, and they leave because they believe there is no reasonable alternative.

Homeless youths are easy prey for criminals. But even when parents report them missing neither the police nor social service agencies will intervene "unless we know he or she is associating with known pimps, or is in obvious need of health care, or has committed a crime" (Ince, *Globe and Mail*, Toronto, Jan. 7 1988: 1). Public agencies frequently avoid intervention because they lack resources, they believe that responsibility must be borne by the individual and by private charities, or because they are wary of being sued.

In Britain social workers have discretionary powers to aid young people until they reach age 21. A few innovative schemes with transitional housing exist. But the lack of any clear statutory obligation, scarce resources, and declining support from central government make these arrangements tenuous. The government, contrary to the recommendations of social workers and its own advisory committee, limits the length of stay for youths in board and lodging. They argue that this policy will compel young people to return home and to reconcile with their parents. Experience demonstrates, however, that this hope is illusory. Usually there is no home available to them.

Social workers estimate that 80,000 young people in Britain are homeless – that is, sleeping rough, squatting, or moving from hostels to shelters or bed and breakfast lodgings. Homeless teenagers have few alternatives to hostels, other than sleeping rough or squatting. They gravitate to hostels after leaving foster homes, an institution, or being required to leave home. Many were evicted from overcrowded, doubled-up arrangements. Some were attracted to the city by job prospects but have been unsuccessful in finding work. A leading cabinet member in the Thatcher government advised young people that the solution to their employment problems was simply to "get on your bike." On the other hand, they are criticized if they move and find themselves homeless, being considered feckless, work-shy, and abusers of the social security system (Saunders 1986: 9).

This is a highly mobile group; half stayed at three or more addresses during the past year. In 1994 Centrepoint, an agency for young homeless people in London, found that half of its residents had slept rough and had no income; 29 per cent had slept in a car; only 2 per cent have full-time jobs (*Housing Association Weekly*, May 27, 1994). Most left home after disputes with parents or foster parents.

Two-thirds of the total were actively searching for other accommodation, usually a bedsit (a studio or bachelor apartment) or cheap flat. They were not interested in another hostel. Virtually all were unhappy with the hostels but they found flats expensive and difficult to locate; they were unable to get on local authority waiting lists and many were excluded by landlords who refused to accept unemployed young people or ethnic minorities.

Hostel staff people are helpful in directing young people to alternative housing but they are unable to provide advice regarding training and jobs. Young people need more than housing; they require counseling, support services, education, training, and employment. Most of the traditional services available for the larger population are not equipped to deal with adolescents on the street: there is a lack of appropriate mental health, substance abuse, crisis intervention, and treatment services (particularly on an outreach basis) for homeless young people who are emotionally disturbed.

Occupancy rates are in excess of 80 per cent for most hostels, and over 90 per cent for places which cater for people who are working or studying. Hostels are provided mainly by housing associations and other voluntary organizations; Community Relations Councils offer places for ethnic minorities; the YMCA and YWCA provide bed spaces for young people working or studying full time; and hostels specifically for ex-offenders and people on probation are furnished by the Probation and After-Care Service as well as voluntary groups such as the National Association for the Care and Resettlement of Offenders.

THE ISSUE OF RACE: NATIVES AND VISIBLE MINORITIES

Minorities are disproportionately represented among homeless people. The U.S Conference of Mayors reported that, in thirteen American cities, 52 per cent of the homeless people are black, 33 per cent white, and 15 per cent other minorities (U.S. Conference of Mayors 1987). A National Institute of Mental Health study in New York, Los Angeles, and St. Louis discovered that racial and ethnic minorities accounted for 65 per cent of homeless people (Martin 1986). It is likely that homelessness will continue to be a significant problem for minorities, in part because they encounter discrimination, but also because they have difficulty securing housing, jobs, and related income support. Established agencies and formal service delivery systems have been found lacking in their ability to reach very low-income minorities (First *et al.* 1988: 120–124).

In the United States African-Americans are over-represented in all categories of homeless people. Though they are very poor, Latinos are under-represented. Apparently, there are several explanations for this situation.

Blacks are subject to more overt discrimination in both housing and labor markets. Latinos have more extensive social and kinship networks which protect them from economic disaster. They are also twice as likely as blacks to be in overcrowded housing; that is, overcrowding is accepted because of the strength of kinship networks and because it is seen as a way to deal with lack of housing and low incomes. Studies of Latinos suggest that "attachment to the personal social network is a key construct distinguishing homeless people from those who are extremely poor but still 'precariously housed'" (Baker 1994: 498).

In Canada Natives (i.e. aboriginals) have high mortality rates. They represent a disproportionate percentage of prison populations, they have a very high incidence of poverty, and a great many are homeless. In certain cities, like Winnipeg, Natives represent about two-thirds of the homeless people using inner city missions and shelters. In 1992 Health and Welfare Canada found that Natives are more than three times as likely to die a violent death before age 65 as non-Natives and twice as likely to die of any cause before 65. The incidence of suicide is 2.5 times higher than the non-Native rate. Native women are more than five times as likely to die from alcoholism or diabetes as non-Native women. Native men die from alcoholism at 2.6 times the rate of non-Native men.

The incidence of overcrowding in Native settlements is about twenty times greater than in non-Native communities. Nearly half of Native housing requires major repair, compared to only 6.7 per cent of the general housing stock. About 38 per cent of Native homes lack running water, indoor toilets, baths, or showers, compared to only 2 per cent in the rest of Canada (Lang-Runtz and Ahem 1987: 45).

A Canadian Bar Association report (1995) shows that, while only 4 per cent of the general population are Natives, they comprise 10 per cent of the males and 13 per cent of the females in prison. Native people, representing 6–7 per cent of the population in Manitoba and Saskatchewan, constitute well over 50 per cent of prisoners. The report concluded that "an Indian youngster in Canada has a better chance of being sent to prison than of completing university." These problems are attributable in part to long-term unemployment and a feeling of powerlessness. Natives, according to the Canadian Bar Association, have been "dispossessed of all but the remnants of what was once their homelands." In 1989 the Canadian Human Rights Commission concluded that, from the justice system to the government to the job market, Natives are systematically ill-treated or ignored (Canadian Human Rights Commission 1989).

In 1991, 81 per cent of Canada's 1 million Aboriginal population was living off reserves. About 70 per cent live in cities. The Aboriginal population of cities increased rapidly during the late 1980s – perhaps reflecting in part a higher level of reporting of Aboriginal origins as well as migration to large urban centers. Natives leave reserves in part because of social conflicts – especially in the case of women and youths – and a lack of economic opportunities. They gravitate to cities, like Vancouver or Winnipeg, where they face a different set of problems. A quarter of Winnipeg's Native population in 1986, for example, had moved to the city since 1981, accounting for half the increase in the number of single parents. Natives in the city are twenty times more likely than non-Natives to receive welfare. Their birth and unemployment rates are two times and three times, respectively, higher than non-Natives. Moreover, their Aboriginal culture is often a casualty of urbanization.

In Britain, as in North America, members of visible minorities often are at a disadvantage in securing affordable housing. During the post-war era new arrivals from the Caribbean and Asia were generally ignored by unsympathetic local authorities. For years they lived in a twilight zone, providing for others like themselves, who were barred from public housing. As blacks and Asians began to qualify for council housing in major cities like Birmingham and London, the local authorities gave them the worst units (usually high-rise systems-built blocks) and concentrated non-whites in slum wards: Islington, Hackney, and Tower Hamlets in London, Toxteth in Liverpool, and Soho in Birmingham. In Soho, for example, three-quarters of the population is in households with New Commonwealth heads (primarily from the Caribbean and the Indian subcontinent).

After investigating the 1981 Brixton riots, Lord Scarman concluded that the most pressing problems are inadequate housing, racial discrimination in the allocation of council dwellings, and lack of comprehensive social and economic policies for urban areas. He found housing to be a major precipitant of violence, calling it one of the "social conditions which create a predisposition towards violent protest" (Scarman 1981: 2.6–2.9).

Profile of an eviction in New York City

On October 29, 1984 a 67-year-old African-American woman, Eleanor Bumpers, was shot to death while resisting eviction from her Bronx apartment. When confronted by several police officers carrying riot shields and weapons, Mrs. Bumpers tried to fend them off with a knife, which she bent on one of the police shields. When the police officers failed to subdue her with a restraining hook, another policeman fired at her with his shotgun. The first shot took off most of her right hand and she immediately dropped the knife. But the officer shot again and killed her. The police did not attempt to use mace or tear gas.

The post-mortem revealed that Mrs. Bumpers, who was $96.85 in arrears on her rent, had never been personally served with eviction papers. Moreover, three weeks before this incident her relatives had tried to pay part of her rent but were refused because the New York City Housing Authority bookkeeper had been instructed not to accept partial rent payments without the supervisor's written consent.

After a three-year delay the police officer who fired the fatal shot came to trial. He was exonerated.

(*New York Times*, November 3, 1984: A27, January 14, 1987: B2)

VETERANS IN THE UNITED STATES

Approximately one in three homeless Americans is a veteran; about one-third of these homeless veterans are from the Vietnam era. Older veterans (over age 54) are disproportionately represented among homeless people. They are more likely to experience problems of alcohol abuse and are homeless longer than other men who frequent shelters. Russell Schutt's survey at Boston's Long Island Shelter found that veterans used the hostel system at twice the rate of non-veteran males (Schutt 1985).

Generally, however, they do not take full advantage of veterans' benefits or alcohol abuse services. A U.S. Department of Labor study discovered that a majority of homeless veterans needed medical, mental health, or alcohol/drug counseling before they could undertake job training and placement. Yet a federal government review of current practices found that "there is no clear-cut responsibility at state or local levels to link Federal and state entitlements and services with private sector homeless services to target veterans for employability and jobs." The Department of Labor concluded that many service providers are unaware of the nature and scope of existing programs for veterans and are not equipped to "address the special characteristics and needs of this hard-to-reach population." Despite the obvious need, there was a 43.5 per cent cutback in staffing for veterans' assistance services during the Reagan years (U.S. Congress 1986: 161). As the veterans of the Vietnam era get older, it will prove to be even more difficult to provide them with appropriate counseling and employment. It is likely that, once they are over age 50, it will be possible simply to maintain them in marginal fashion.

Vietnam veterans

Ken Smith of the New England Shelter for Homeless Veterans in Boston noted that the average age of men he sees is 44; most veterans are now close to that age:

> most are veterans of Vietnam who are suffering from post-traumatic stress syndrome which, combined with homelessness, renders them unable to cope; many find themselves living on the streets, foraging for food . . . They are disenfranchised from their families, chemically addicted to alcohol or drugs, very lonely, isolated for long periods of time . . . Some of them just need a safe place to re-group and restructure, but it's difficult to find a safe place.
>
> (Ken Smith interview on "As It Happens," Toronto; Canadian Broadcasting Corporation, December 2, 1993)

PEOPLE WITH DISABILITIES

James Wright's study of health clinics for the homeless in American cities found that "about one client in six is physically disabled and incapable of working" (only half of them, however, receive disability benefits) (Wright 1988–89: 23). Persons with disabilities make up between 10 and 15 per cent of the population in Britain, the United States, and Canada. Roughly one-third of all disabled people are age 65 or older. At least half of the total have a moderate or severe disability and live at or below the poverty line. A Canadian government survey found that 63 per cent of the country's 1.8 million adults with physical and mental disabilities have poverty-level incomes. Fewer than 40 per cent are employed (Canadian Association for Community Living 1989).

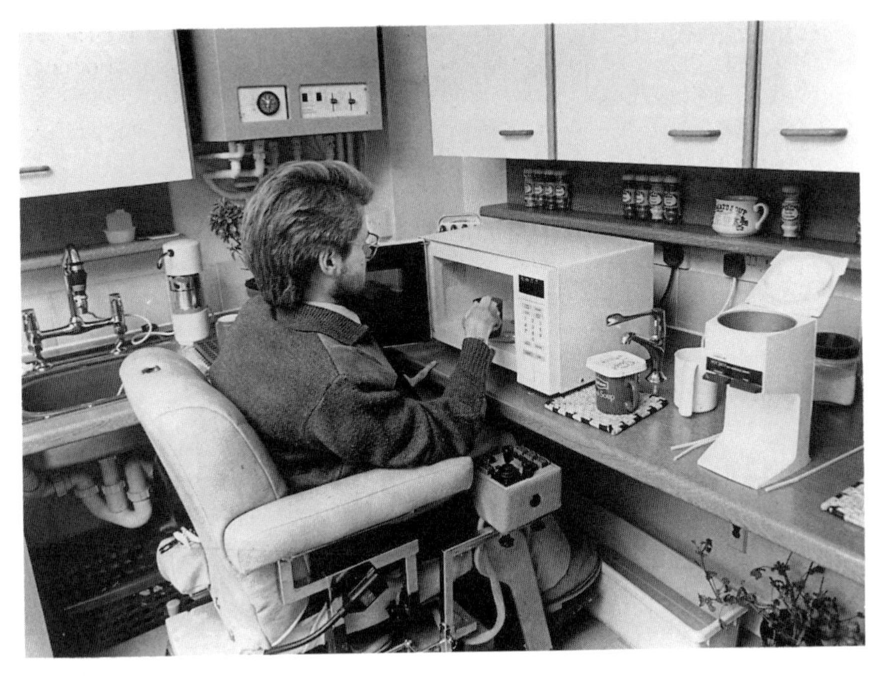

34 Richard Shaw, age 31, has muscular dystrophy. He now lives in housing specifically adapted for him by the Centre for Independent Living in Derbyshire, England. This move, however, caused him to lose grant money (because he vacated his parents' home) and to run afoul of the social service establishment, which he found to be unsympathetic to his desire to live independently
(*Roof*, March/April 1990:27; photograph by Brenda Prince, courtesy of Shelter)

Many individuals with disabilities are homeless or vulnerable; they have higher rates of unemployment than people without disabilities, and tend to have lower incomes when they are employed. For some, particularly in the United States, a significant portion of their earnings must be spent on care and medication. A recent study in Britain found that:

> services are chaotic, fragmented and often fail to deliver appropriate care. In many places, severely disabled people are forced to choose between living at home with little or no help from statutory services, or being institutionalized. This all or nothing approach places heavy burdens on people with physical disabilities and their families.
>
> (Beardshaw 1988)

THE FRAIL ELDERLY

In all three countries the proportion of the population made up of people over age 64 will increase dramatically over the next two decades. In Canada there were 2.7 million older people in 1976; this will rise to 3.9 million in 2001 and to 6 million in 2021. By the year 2000 older people will represent 12.2 per cent of the American population, 12.8 per cent of Canadians, and 14.5 per cent of Britons. In some regions the population over age 75 will virtually double. Many are vulnerable to homelessness and four in ten are disabled (CMHC 1989: 8–13). Older people typically have low fixed incomes and have difficulty maintaining dwellings in the face of inflation and increasing maintenance expenses and rents. A pensioner in England describes the day-to-day reality for older persons in marginal circumstances:

> When you are poor, you have to be extra good at managing, extra good at self-discipline, extra good at economising, extra good at managing without and extra good at dealing with constant crisis, stress and frustration. Few of us are extra good. The rich aren't.
>
> (Campaign Against Poverty [Shelter] 1989)

These problems are exacerbated by the deaths of friends and spouses, declining health which may reduce their mobility and sense of independence, and by social isolation. In these circumstances the home and neighborhood assume added importance (Rapoport 1985: 255–286). For those whose mobility is restricted the quality of their home environment will determine in large part the extent to which they are able to maintain their independence.

Most of these older people are women. Women generally marry at an earlier age than men, they marry men who are older than themselves, and they tend to live longer than their husbands. In Britain women outnumber men by a 2:1 ratio in the 60–65 age group; in the growing category of age 85+ there are 3.25 times as many women as men. Over 60 per cent of women over age 75 live alone, while only 29 per cent of men in this age category are alone. About 40 per cent of older people are poor – most of them are women – and many also live in inadequate housing: there are now 1.3 million elderly people in poor housing (Sykes 1994: 77–81).

REFUGEES AND NEW IMMIGRANTS

Applications from immigrants and refugees who need housing and social services have been increasing in all three countries. In Canada, for example, immigration increased 2.5 times between 1985 and 1991, to an annual total of 218,000; two-thirds of these newcomers gravitate to the three largest Census Metropolitan Areas (Toronto, Montreal, and Vancouver). They face extraordinary difficulties in adjusting to new customs, climates and languages. Some families, for example, are forced to separate when fleeing oppressive regimes; others find themselves in large North American or British cities in winter-time, having just arrived from a rural area of tropical countries in Central America, Africa, Asia, or the Caribbean.

Recent immigrants are exploited by employers. Women in the garment industry, for example, or those employed in jewellery factories, regularly work 60-hour weeks at minimum wage without overtime pay or benefits and without coverage by Workers' Compensation. Employees are usually ignorant of the laws; most speak little English. In any case, they are reluctant to bring a complaint against their employer for fear of losing their jobs or because they are "clandestines" who can be deported. These precarious jobs are usually passed to immigrants by friends or relatives.

New immigrants face discrimination in the housing market as a result of language problems and lack of knowledge of their rights. The uncertain immigration status of some refugees means that they lack access to government-assisted housing and welfare benefits. Recent arrivals frequently live in apartments owned by other immigrants; both owners and tenants are accustomed to overcrowded and inadequately maintained dwellings. Even when living in very poor housing, they are reluctant to express dissatisfaction for fear of being evicted. In order to gain access to suitable housing and to secure employment they may require orientation and support services, literacy programs, and legal aid.

ROOMERS

Many think of roomers as transients. In fact most live in a particular rooming house for a year or longer and then move to another room in the same district. A Toronto survey found that about three-quarters are men and most (more than three out of four) are over the age of 35. As a group they have little formal education – many are illiterate. Less than one-third are currently employed and another third rely on pensions as their main source of income.

In Vancouver's Downtown Eastside there are 10,000 people, most of whom are single men living in SROs and rooming houses. Their average age is 57 and over half have some form of disability, usually from having worked in logging, fishing, or as longshoremen. Virtually all are on low fixed incomes, and about 60 per cent receive social assistance. Housing consists of rooms which are about 10 feet by 12 feet in area; 70 per cent have no cooking facilities, and 90 per cent lack a private washroom. The buildings are old and poorly maintained.[2]

All but a few of these individuals are living in poverty. More than seven in ten roomers pay more than 25 per cent of their income on rent and roughly three in ten spend more than half of their income on rent. Half pay their rent weekly and only 2.8 per cent have a written lease. About 70 per cent live alone in a single furnished room. Only one in five roomers is able to control the temperature within the room and one in four find their units uncomfortably hot in summer and cold in winter. Most complain that their rooming houses are infested with mice and cockroaches.

One-third of the roomers have no close friends within the same rooming house. The principal problems encountered by these people are with their accommodation, with lack of money (18 per cent), with loneliness and physical health. Well over half have been ill within the last 12 months. About three-quarters of the roomers wish to move, citing the poor quality of their furnishings and facilities and in particular the lack of privacy, space, freedom, and accessibility. Most want to find a self-contained apartment, rather than simply a unit in a rooming house. Seven in ten who move, however, remain within the rooming house network.

Sixty per cent of the respondents indicate that they live "only for today," they hold out little hope for improved conditions, and most cannot conceive of fighting for better legal protection and guarantees of civil rights for roomers.

In Ontario roomers are now protected by the Landlord and Tenant Act and the Rent Control Act, but in many other jurisdictions they are covered only by "innkeeper's" legislation which is silent on rights for residents; they can be evicted without notice, without cause, at any time. Rents can be raised by any amount at any time. There is no mechanism to deal with disputes, property can be seized, and mail will not be forwarded by landlords or by the post office (McMaster and Browne 1973).

In some rooming houses many of the residents are former psychiatric patients and alcoholics. In January 1989, a resident of the Rupert Hotel in Toronto ignited some garbage in one of the rooms, creating a blaze in which ten people died. Since then, inspections have become more thorough, but most rooming houses still fail to meet minimal building and public health standards.

THE CHRONICALLY HOMELESS

Those who are without any housing, who sleep rough or use emergency shelters, represent the bottom end of a continuum of people at risk of becoming absolutely homeless. The number of chronically homeless individuals varies significantly over time and from place to place; a rough estimate is that this group is about one-third the size of the population which is vulnerable or "at risk." Chronically homeless people are distinguished, not by what they have, but by what they lack. Individuals who are classified as "hard to house" – a label which some find problematic – have little income and virtually no housing options. They spend from 50 to 80 per cent of their meager resources on shelter.

In addition to a range of affordable housing options, these individuals need a variety of community support services to enable them to achieve stability. These may include community mental health services, life skills training, budgeting and money management, drug and alcohol programs, crisis intervention services, peer support groups, literacy education and employment/ vocational assistance.

Roomers

In one rooming house, near the Queen Street Mental Health Centre in Toronto's West End, a dingy three-storey building contains fifty-one units, renting for $350 per month. Rooms are about 10 feet wide and are sparsely furnished:

> In one of the rooms . . . a side table serves as an ashtray, its top covered in cigarette butts. Blue pills lie scattered about near the door on the grimy floor. One of the burners on a greasy two-element stove is turned on high . . . The tenant says the burner is the room's only source of heat. He also says he has to share a washroom with residents of seventeen other rooms on the floor. In October 1993, the City of Toronto refused to renew the building's rooming-house license after the landlord failed to comply with a dozen work orders issued over the previous year. By January 1994, however, the rooming house was still occupied as the residents awaited a court hearing.
>
> (McLeod, *Toronto Star*, October 11, 1993: A14)

Street people frequently need help on an episodic basis. Institutional service providers, however, generally have standardized packages which require participation on a regular, long-term basis. This approach fails to meet the needs of many homeless people, who feel that the key to escaping the cycle of dependency is support services which allow the individual to become independent. Flexibility is essential. With adequate support in coping with problems themselves, they become less dependent on social workers and other professionals. The problem with the current system of service provision is not so much the level of service as the method of delivery and the assumptions or values on which delivery is based.

Homeless people and advocates

Issues of power and control are crucial to homeless persons. Leon Zecha of the San Francisco Homeless Caucus stressed this point in his testimony to Congress:

> We need to organize because social workers and other professionals working on our behalf, no matter how well-meaning, do not understand our needs. For instance, some of our advocates, representatives of major provider agencies in San Francisco, in a proposal to the city on managing hotels for the homeless, wanted to remove all the doors from the rooms and separate couples because genders must not be mixed.
>
> (U.S. Congress 1984: 342)

In the past, shelter, free meals, and services were often provided in the same location. Often, an individual was compelled to enroll in a rehabilitation program in order to qualify for housing. This approach generally has not been successful. A recent trend has been to separate housing and support services, in order to avoid the stigma of "dedicated projects" which deal solely with a particular marginalized population, like alcoholic men. A number of successful models include housing, whether in shared units or private apartments, which offers support based on the concept of facilitation rather than compulsion. The facilitative model provides services in adaptable fashion. Support is available for those who require it, but a conscientious effort is made to foster independence rather than dependency. Individual residents participate in the planning and design of the building and living units, and deal on a continuing basis with such issues as maintenance and screening applicants for vacant units (Single Displaced Persons Project 1987).

As noted in the previous chapter, a substantial percentage of people who are deinstitutionalized from hospitals and prisons become homeless, often reaching a chronic state in a brief time because they lack social supports. NACRO in England (the National Association for Care and Resettlement of Offenders) found that up to 90 per cent of prisoners had no permanent housing available at the time of their release.

In the United States, where the rate of imprisonment is more than four times higher than Canada or Britain, recent trends in incarceration raise troubling questions about possible effects on chronic homelessness in the future. There are now more than 4 million Americans in prison or on parole;

94 per cent are males; two-thirds have not completed high school; half are African-American or Hispanic; two-thirds were convicted of non-violent crimes (mainly drugs and property offenses). Though the murder rate remained level during the 1980s, the U.S. prison population doubled and governments spent over $20 billion on this system in 1990; new prisons are being constructed at an average cost of $60,000 per cell.

RURAL HOMELESSNESS

Homelessness is not just an urban issue. Rural areas in the United States have been severely affected by recessions. Many small farmers have lost their livelihoods as a result of economic trends in agriculture: mechanization, which requires huge outlays for equipment and interest payments, favors large corporate farms which can achieve "economies of scale." Rural regions have lost population as a result of aging and out-migration plus economic decline. Public tax bases, as a consequence, have declined, and rural counties are unable to provide sufficient social services to meet the needs of people who are vulnerable to homelessness.

Although it is generally assumed that rural homelessness in Canada is non-existent because of climatic conditions, there are many rural people who are homeless or vulnerable because of fires, domestic violence, mortgage foreclosures or evictions from farms and homes. Many live in inadequate housing; one out of four rural dwellings is deficient. Home to about 6 million Canadians, rural districts are twice as likely as cities to have substandard housing (Statistics Canada 1990). When rural people encounter a housing problem they have few alternatives, short of doubling up or moving out of the community. No emergency shelters are available except in cities. Often they lack the funds to undertake needed repairs or they live in areas where only a marginal cash economy exists.

In much of northern Canada little or no private housing is available. The northern housing situation is characterized by an inhospitable climate (which means that houses are expensive to build and maintain), transportation difficulties (materials can be shipped in to some communities only once or twice a year on barges), and poor social and economic conditions. Unemployment, substance abuse, domestic violence, and suicide are common.[3] Severe over-crowding is the norm in these remote regions. In some places as many as sixteen persons live in three-bedroom houses, utilizing closets and furnace rooms for sleeping. When people are forced to live indoors for prolonged periods, it is almost inevitable that social and domestic problems will worsen.

SUMMARY

While most homeless people have certain things in common, like health problems, low incomes, and inadequate housing, there are significant differences among the groups considered at risk. Women, for example, are subject to physical and sexual abuse, and those on the streets must fear for their safety. Over half of single mothers have serious housing problems. Children

Communication

I have learned that the biggest problem for the extremely poor is one of communication. The educated, power elite don't have the time to wade through "poor" language skills and the poor don't understand "specialized" languages.

For the first three years of my involvement, I was frequently embarrassed by the bureaucrats. I would attend a housing conference, a health and human services conference, a training seminar and people would make fun of my lack of teeth, or poor clothing. I learned to feel sorry for them, and to learn in spite of them. I would have thought they would encourage me.

I persevere. Most of the power for housing issues lies in housing authorities and home ownership programs. I hope I have learned enough of how these programs work . . . to apply them to homelessness. I have also been learning community organizing techniques during the past five years. Through Oregon Fair Share I have been given access to some excellent opportunities.

So now I must overcome the personal inhibitions, the feelings of low self-esteem and the fear-of-success syndromes of my past.

(John Statler, a formerly homeless man, Medford, Oregon, May 1995)

are in jeopardy; in addition to illnesses, they have a high incidence of learning disorders, emotional difficulties, and problems coping with school. As they reach their teens young people on the streets fall prey to pimps, pornographers, and drug dealers. With little education and no skills they have virtually no hope for a better future. People with disabilities, both mental and physical, are over-represented in the homeless population. Other high-risk groups are frail older people on low fixed incomes, refugees and new immigrants, Natives and other visible minorities, and individuals who have been deinstitutionalized. People from these groups are not always well served by existing social service arrangements. There is not so much a lack of services as inappropriate or inflexible delivery mechanisms. For traditional service providers it has been difficult to adapt services to the needs of these people who are considered hard to reach.

8

MORE THAN JUST A ROOF

> What is the use of a house if you haven't got a tolerable planet to put it on?
>
> (Thoreau)

> I want my own place . . . a space to have family or friends; privacy – especially to have my own bathroom . . . A home means comfort, rest, a lock on the door; it's mine to come home to and close the door.
>
> (Crane 1990: 40)

THE SIGNIFICANCE OF HOME

Vaclav Havel believes that all people must have room to "realize themselves freely as human beings, to exercise their identities." Our homes are "an inseparable element of our human identity. Deprived of all the aspects of his home, man would be deprived of himself, of his humanity" (Havel 1992: 30–31). In addition to providing refuge from the elements, a dwelling offers physical and psychological security. The home connotes status, helping to define our social position. It represents a source of pride, self-respect, and a tangible measure of our economic worth. For some, the family home provides roots in the community and a link with previous generations. The home as haven offers privacy, respite, a peaceful setting for regaining physical health and maintaining mental well-being. The notions of home and family are intertwined:

> "Home" brought together the meanings of house and of household, of dwelling and of refuge, of ownership and of affection . . . This wonderful word, "home" . . . connotes a physical "place" but also has the more abstract sense of a "state of being".
>
> (Rybczynski 1986: 66)

Homes allow us to introduce a sense of order in our lives. For most of us, our home offers convenience and a comfortable environment in terms of temperature, ventilation, lighting, and soundproofing; it provides a fixed address, where we can prepare meals, receive mail, take phone calls, greet friends, and store belongings without outside interference. The home is not confined to four walls. Inside and outside are connected. We derive satisfaction from the use of adjacent outdoor spaces – gardens, courtyards, lawns, balconies, decks, or porches – which allow us to remain close to our house, to keep our eyes on the street, while socializing with neighbors.

Self and circumstance are inseparable in the home (Dovey 1985). Homes are a mirror, a reflection of our state of being and an extension of our human presence. The home constitutes physical, social, cultural, and psychological space which, on the one hand, shapes our behavior and, on the other, helps to form our perspective on the world. The dwelling, then, can enlarge our world view, can reinforce our sense of self-esteem; conversely, it can confine us, increase our sense of insecurity, and severely constrict our freedom and our horizons.

Almost home

In NYC,
- a psychiatric hospital bed costs $113,000 a year
- a prison cell costs $60,000 a year
- a shelter cot costs $20,000 a year
- a permanent home and supportive services costs $12,500 a year
Which would you invest in?

This is the headline of a full page advertisement placed in the *New York Times* (May 28, 1995) by Almost Home, a coalition of business people who advocate supportive housing and oppose a proposal before Congress to cut shelter programs.

Almost Home points out that "combining affordable housing with appropriate services, including help in finding work has consistently succeeded in helping people get off the streets and rebuild their lives." They contend that "Supportive Housing is not only more effective, it is far less expensive than traditional responses to homelessness. Shelters, hospitalization and legal interventions consume our tax dollars without permanently reducing homelessness."

Moreover, home cannot be adequately described without reference to the larger context in which that home is situated. Thus, home is tied to our notions of community and to the process of commodification which has come to define home in modern terms. In the early post-war years these ideas were exemplified in the racially exclusive and economically homogeneous suburbs which emerged from fields on the urban fringe. Later, similar concepts were embodied in the revealing term "starter home," which describes the first step in an inexorable process of accumulation. The long struggle to create livable public housing in North America has more to do with its economic spin-offs and the political need for social control than with a desire to house poor people. In Western societies, home is a social creation, rooted in economic ideology, and made possible (for those who can afford it) by political and legal institutions. Residential development is a spatial process that defines the world in which we dwell; it assigns use- and exchange-values to these spaces and, in turn, to the people who inhabit this built environment (Bresalier and O'Donnell 1995: 3). The way homes are provided, for whom, and by whom, is revealing of the extremely close and enduring ties between business and government. It also offers some insight into the power and class differentials and the differential nature of expectations of what constitutes a suitable dwelling for various populations.

For some, this means that their living circumstances offer neither haven nor refuge. The cases which follow illustrate that the reality of life for those without secure housing is far removed from the comforting images conjured up at the beginning of this chapter, because the temporary lodging provided to homeless people lacks the social, emotional, and psychological attributes of a real home.

CASE STUDIES: B & BS, WELFARE HOTELS, EMERGENCY SHELTERS, AND WOMEN'S REFUGES

The first case is based on recent experience with bed and breakfast hotels (B & Bs) in England as temporary accommodation for homeless families. The second describes the American response, using the example of welfare hotels in New York City. The third deals with hostels or emergency shelters, using Toronto to demonstrate the futility of employing such temporary expedients to respond to the long-term problems associated with homelessness. The final case describes a variety of refuges for women in Britain and North America.

Bed and breakfast hotels in England

At the end of 1994, in England alone, there were more than 50,000 households in different types of temporary accommodations, including hostels and bed and breakfast (B & B) establishments, up from 5,000 in 1982. During the 1980s the B & B became a common expedient for dealing with homelessness, particularly in London's inner boroughs. Critics point out, however, that this is neither short-term, nor a solution. Though referred to as hotels, most take in only low-income people placed by local authorities; most serve no meals. In a well-publicized case the court found it "astonishing" that the local authority could regard it as reasonable for a family of seven to live in one room measuring 10 by 12 feet (*R v. Westminster City Council ex parte Ali 1983*).

The government's review of B & Bs and houses in multiple occupation (HMO) found that 300,000 (81 per cent) lacked proper means of escape from fires; 220,000 (61 per cent) did not have essential amenities; 59 per cent were unfit or in disrepair; and 4 per cent were in such deplorable shape that condemnation was recommended. Surveyors in London discovered that 62 per cent of hotels used to shelter homeless families in London would fail the required environmental health inspection (*Roof*, July/August 1987). They concluded that conditions in these establishments are, in effect, a prescription for poor health. Facilities for storing, preparing, and cooking food are extremely limited or non-existent. The Association of London Authorities' code of practice for B & Bs specifies that full kitchen amenities must be available for every five residents. In practice, however, few places meet these standards.

Often the hotel room is on one floor, the WC or toilet on another, the bath or shower on still another floor, and the kitchen, if available, is in the basement. This layout makes life especially difficult for single mothers. They find themselves in a situation where they have to lock some of their youngsters in their room while they use the WC or kitchen.

35 Young couple in hostel with 9-day-old baby, Middlesbrough, England, 1990 (Photograph by James A. Gardiner, courtesy of Shelter)

B & B living

"I walked into B & B with a partner, but I walked out a single parent. It causes it. When we were in B & B we didn't have a job – no one'll give you a job without a proper address. Since I've got the flat I've got a little part-time job. I can get a bank account. I can do anything I want because I've got an address. That's the most important thing for someone to have."

"I am treated differently now. You sometimes came across people who, when you said you were living in B & B, gave you that look. It's difficult to describe but you feel it. They make you feel small and you're nothing. Now that you have your own place you think, I can start living normally like other people. I can start sorting out my life. You start to think ahead."

"When I was rehoused I just felt really relieved. I thought it was going to be my third Christmas in B & B, and Christmas in B & B is bad."

(Crane 1990: 40)

Maintenance and repairs are poor, resulting in hazards, particularly for young children. In a number of cases mothers and their infants have died in fires which engulfed poorly equipped B & Bs. A substantial percentage of these places lack fire alarms and telephones for emergency calls. A health visitor in London submitted the following report on hotel conditions:

> I went to do a new birth visit in a bedsit [a one-room apartment], where there was a young woman and her baby and five other single women.

> Some cowboys [i.e. fly-by-night contractors] had put in an electric meter that was smoking. When the residents complained, the landlord gave them a fire extinguisher and told them not to use it unless they really had to, as the insurance would cover any damage. I phoned the environmental health who said they had a backlog of work and were not going to act very quickly. So I made an anonymous phone call to the fire brigade who then made the environmental health close the basement down.
>
> (Conway 1988: 12)

Those who are forced to use such hotels are already on the margins of society, usually unemployed and living on low incomes. Two-thirds of the households in B & Bs are headed by women. Many are immigrants who confront language or cultural barriers, making it more difficult to deal with a complex bureaucratic system. Residents usually have no idea how long they will be confined to B & Bs. The average length of stay, while waiting for a vacancy in council housing, is about two years.

> The biggest difference is privacy and feeling secure. There was always a fear, an uncertainty of whether you were going to get a place or not when you were in B & B. You're never informed what's going on, from one week to the next, from one month to the next. Once you're put in B & B no one bothers to contact you at all. They put you there and seem to forget about you until your name comes up next on the list.
>
> (Crane 1990: 40)

A London doctor whose practice includes a number of B & Bs reported that residents have "more psychiatric problems, depression, anxiety and sleeplessness, and homeless children suffer from more upper respiratory tract infections . . . Gastroenteritis is common and we have had an outbreak of measles and chicken pox" (Conway 1988: 43).

The government's Audit Commission (1989: para. 22) found that "B & B hotels usually offer the lowest standards at the highest costs." In 1994 the Housing Minister, Sir George Young, conceded that "the use of bed and breakfast is now almost a dead issue . . . Government policy is quite clear. B & B is an unsatisfactory form of accommodation."[1] Many local authorities responded by shifting from B & Bs to leased accommodation in order to save money and to improve the situation of homeless people. In general, this represents a considerable improvement because families now have more privacy; generally, the flats are self-contained with their own kitchen and washrooms. Nevertheless, in 1994 local authorities were still booking 2.2 million nights in these places at a cost of £65 million (*Roof*, July/August 1994: 23).

New York's welfare hotels

Officials in North American cities have employed similar stopgap measures. The most notorious are New York's welfare hotels. Before Ed Koch became Mayor of New York City he ridiculed John Lindsay for sheltering families in

36 Resident of B & B, London (Photograph courtesy of Shelter)

welfare hotels. After becoming mayor, Koch tripled the number of these accommodations and spent $100 million annually on welfare hotels at the end of his last term in office. Numerous exposés documented the conditions in these hotels, some of which are owned by prominent contributors to city political campaigns. One property, the Jamaica Arms, was purchased from the municipality in 1982 for $75,000; four years later the city paid the owners $1.2 million for the temporary use of their rooms. Another, the Holland, was allowed to continue operating while the property had 1,000 housing code violations outstanding. Ironically, many of the places used as welfare hotels were hastily converted from SRO establishments, after the tenants had been evicted, in order to reap the benefits of the government's homeless aid program (Kozol 1988). New York City's J-51 tax law helped to finance the conversion of 535,000 SRO units between 1978 and 1985.[2]

At the time taxpayers' dollars were being paid to hotel operators, the City of New York owned 100,000 abandoned apartments. Because of restrictions on federal funding, however, it was difficult to rehabilitate these units or to secure funds for new construction. It was speculated that the atrocious conditions in welfare hotels were being tolerated by bureaucrats in order to discourage families from becoming homeless and thus qualifying for public housing.

Life in welfare hotels

... the government spends $37,000 a year to house a family of six. Once food, social services, and administration costs are factored in, the family's effective income is well over twice the national median. But due to the inherent inefficiency of the state, they are crowded into vermin-ridden rooms with no cooking facilities, where they live on hot dogs warmed under the tap. Prostitutes ply their trade in adjacent rooms and drug lords are eager baby sitters. Kept in such conditions for an average of a year and a half, the families swiftly deteriorate: marriages break up, the work ethic erodes, and drug habits are learned.

(Coulson 1987: 15–16)

Eighty per cent of the occupants of these hotels were on welfare before becoming homeless; 85 per cent of the households are headed by single women, most of whom have never worked. The great majority are non-white, illiterate, and ill-informed about their rights.

It was not until the federal government signaled its intention to drastically cut its reimbursements under the Emergency Assistance Program that the City of New York decided to vacate the welfare hotels. By the end of 1990 the number of families in the hotels had been reduced to 650, less than one-fifth the level of the late 1980s. Most of the homeless families from these hotels were moved to vacant public housing units. This, however, created a backlash among households on waiting lists or doubled-up in public housing projects. Subsequently, the number of families in welfare hotels rose again despite efforts by municipal officials to discontinue this practice (Blau 1992: 161–162). New York's welfare hotels are perhaps the most blatant example of public policy on homelessness run amok. But similar programs exist in many areas of the country where housing is in short supply.

Hostels and emergency shelters in Toronto

It is increasingly well understood that hostels are not an adequate response to the present lack of long-term housing for low-income singles. However, it is less well understood that the present functioning of hostels reinforces the homelessness of the people who use them.

(Single Displaced Persons Project 1983)

A ubiquitous feature of the urban landscape is the emergency shelter. These establishments are typically large, formerly vacant buildings, ranging from warehouses to gymnasiums. Although some are disreputable barracks where older alcoholics and families are housed together, not all are Dickensian. Some shelters serve an important emergency function. The dilemma posed by this temporary expedient, however, is that it is becoming a permanent fixture in urban society.

Emergency shelters or hostels for single men have been used in North American cities since the nineteenth century. Starting in the 1970s, however, emergency or transitional shelters for families, women, and youth appeared. But it was not until the mid-1980s that they became ubiquitous in cities.

37 Friendship Drop-in Centre, Toronto, 1988 (Photograph by Byron J. Bignell)

These shelters supply basic accommodation on an emergency, time-limited basis for those who lack other options; in some cases they provide a transition to more permanent and secure housing.

Municipal shelters are operated in most urban areas. Local governments sometimes contract with private agencies to staff and administer them. Cities like Washington, New York, and Chicago had to be pressured by such advocacy groups as the Coalition for the Homeless and the Low-Income Housing Coalition before they reluctantly became involved. Civic leaders generally resisted these demands, fearing that shelters would attract migrants and transients who are viewed as an eyesore and a detriment to business. City shelters are often large, bureaucratic and expensive: New York's largest consists of 1,200 cots set out on the floor of an immense armory.

Toronto's Seaton House, a shelter for almost 700 single men, provides 326 beds for unemployable permanent residents. The remainder of its spaces are used by employable, usually younger men. While the shelter does provide refuge from the frigid winter winds, its air quality is poor. Respiratory and pulmonary disorders, colds, and the flu, are readily passed from one to another in the cramped quarters.

The superintendent of Seaton House in the late 1980s reported on the philosophy underlying the shelter's operation:

> All we are doing is keeping them alive. We're feeding them, giving them a place to sleep, but we're not solving their problems . . . I'm not suggesting that it is a realistic objective of a place like Seaton House to try to solve the problems of unemployment. We're not in that business.
>
> (*Now Magazine*, Toronto, January 30–February 5, 1986)

Toronto's Shelter for Men

The iron bunks are about two feet apart. The upper ones are high; one man tried three times to swing himself up before he made it. Over 200 men were in the late arrival section where I was. Most of them were young – many under 25. There were ex-psychiatric patients. Released into a community without support systems, they end up here; some of them cry out in the night . . . Badly drunk men are put on floor mattresses and left alone . . . The stench of sweat, smoke, vomit, urine and unwashed socks was overpowering . . . At 5:15 a.m., the lights come on. The man beside me said it's the worst time of day: "You sit on the edge of your bed and wonder what's to become of your life, and if anybody cares."

(McLeod, *Toronto Star*, February 10, 1987: A13)

Not surprisingly, some of the wintertime occupants of Seaton House opt to sleep in a nearby park once Spring arrives. One resident described the effect of having to rely on this shelter of last resort:

> Before I entered this place I was feeling okay. Not that great, but not too bad. The moment I walked through the door I could feel something, tension or something that was totally different from the outside. It made me feel different too, edgy and nervous. Like I had no idea what would happen next . . . I'm not like the rest of these guys. These guys are the scum of the earth. They are lazy, they don't give a damn about themselves or anybody else. I'm here because I have to be here. There is nowhere else to go. If there was I'd go there. As soon as I can get back on my feet, get things together, I'll be out of here. If I thought I'd be here the rest of my life, I'd get in front of the next streetcar.
>
> (Garrow 1986: 114)

The policies and practices of hostel operators (in some cases) help to ensure their use as long-term shelter by failing to provide a setting where people can re-establish themselves, mentally and physically, before finding permanent accommodation. These practices include placing a large number of residents in a dormitory-type setting; inappropriate short-term limitations on length of stay (which tend to induce transience);[3] night use only (residents are forced to leave early in the morning); minimal staff-to-resident ratios (which makes it difficult to provide a secure environment); staff members who adopt a controlling or punitive approach to residents; limited resident involvement in the planning of the shelter or its operations; and only tenuous connections to supportive community services. Many of the shelters offer no meals, no personal space for storage or privacy, very limited common areas, scant attention to the hygiene of residents, and primitive sleeping arrangements – in some places residents must sleep on floor mats.

These practices tend to induce feelings of shame, inadequacy, and hostility. One resident of Seaton House described his feelings at having to rely on the shelter of last resort:

> Let's face it. I'm a loser. Everybody in here is a loser. You think if we weren't losers we'd be here? If we weren't we'd find some way to get out of here. My father survived the Depression and the War. You think I could look him in the eye if he knew where I was?

Punitive policies do not serve the desired purpose. The net effect might be to discourage use of hostels if residents had any alternative. But most have nowhere else to turn. After short stays, people often leave without having assembled the necessary supports to sustain them. The shelter experience is part of a downward spiral which has a negative effect on the individual's personal appearance, morale, and physical and mental well-being. When a person becomes desperate enough to rely on hostels it usually means that he or she has exhausted all other options. One of the disturbing observations of current hostel operators is that a younger generation is becoming dependent on shelters as a quasi-permanent form of housing.

There is a continual deterioration in their situation – physical or mental health, stability of friends or family, legal problems, financial and emotional independence, job prospects. Thus, the lengths of stay increase, the periods between stays shorten and eventually a hostel, or a string of hostels, becomes their permanent housing (Single Displaced Persons Project 1983: 9).

On the night of January 22, 1987 the Canadian Council on Social Development conducted a "snapshot survey" of shelters and reported that Canada had "472 facilities (with 13,797 beds) that exist primarily to serve the homeless and destitute" (Canadian Council on Social Development 1987: 1–4). More than 100,000 different people used these shelters at some time throughout the year, representing about 4 persons per thousand in Canada. Table 7 describes shelter users: 61 per cent are men, women represent 27.5 per cent, and children (age 15 and under) make up the remaining 11.5 per cent.

A profile of Drina Joubert

Just before Christmas, 1985, Drina Joubert's frozen body was found in an alley in downtown Toronto. She was one of about forty homeless people who freeze to death each year in Ontario, one of the wealthiest regions of Canada. In the months before she died Drina Joubert had been refused entry to a number of hostels in Toronto, had been beaten and robbed by three young men, and had been through the revolving door of detox institutions many times for her alcoholism, mental illness, and physical health problems.

According to the Coroner's Jury, "she sought help from practically every available social agency and hospital service in the City of Toronto." The Jury concluded that "the bureaucracy designed to help the most disadvantaged among us has become unresponsive to the need of people it was created to serve. It is fragmented and inefficient."

The manager of a downtown hostel testified at her inquest that Drina Joubert "was becoming more and more disturbed and angry . . . and posed a danger to other residents . . . We don't have facilities and staff to deal with such difficult people."

In Toronto, which has the highest incidence of homelessness in the country, more than 26,000 different individuals use thirty emergency shelters annually, plus another ten temporary shelters, including motels (Municipality of Metropolitan Toronto 1995). Use of the hostels increased from a volume of 694,000 in 1990 to more than 1 million person days in 1995 (one unit equals

Table 7 Characteristics of emergency shelter residents in Canada

Situation	*% of sample*
Unemployed	54.7
Receiving social assistance	51.5
Alcohol abuser	33.3
Current or ex-psychiatric patient	20.1
Drug abuser	15.0
Evicted	9.4
Physically handicapped	3.1

Source: Canadian Council on Social Development (1987)

the use of a hostel for one day by one person). Average length of stay is 30.6 days, but most families and battered women stay longer than a month, while younger people and most of the adult men stay for shorter periods. The average person used the shelter system 2.4 times during the year. Fifty-eight per cent of shelter users are men, two-thirds of them over the age of 24. Fifteen per cent are children; over half of them under age 6; one-third are ages 6–12, and 10 per cent are 13 years or older. Women alone represent 15 per cent of shelter users. Excluding children, the male/female ratio is 3:1, reflecting the nature of existing facilities. Table 8 shows that over half of the shelter beds are designated for single men, and most of these are in very large shelters (Municipality of Metropolitan Toronto 1993, 1995). In the first half of the 1990s while use by adult men decreased slightly, there were substantial increases in shelter use by families and by battered women. The Director of hostel services commented that "we are able to accommodate these high numbers [of families] through the motel arrangements we have in place. Historically, the occupancy levels for families rise during the summer months, so it's anybody's guess as to what lies ahead." He was most concerned about accommodating adult women: "this is a serious concern since there are only 257 beds available and occupancy is now in excess of that level. The eight shelters that serve adult women simply cannot handle any further increases in demand" (Municipality of Metropolitan Toronto 1995: 1–2).

Table 8 Capacity and characteristics of Toronto shelters

Shelter type	*1982*	*1987*	*1993*	*No.*	*Aver. size*
Single women	19	247	257	8	32 beds
Single men	1054	1357	1357	11	123
Youth	76	185	261	6	43
Families	154	539	861	5	172
Total	1303	2328	2736	30	91

Source: Municipality of Metropolitan Toronto (1995: 1–2)

Tough choices in running a shelter

"Running a shelter is a great deal different from thinking about running a shelter; and being executive director of a shelter is some other animal than sitting on its Board of Directors developing 'operational' philosophy. You who reject out of hand the concept, and certainly the practice, of choosing among the homeless regarding the provision of services, maybe you have some magic solution to situations like the following . . .

- What do you do with the guest who insists, every time admitted to the shelter, on endangering other guests?
- What do you do when a man shows up at night at the shelter in sub-freezing weather with his 12-year-old son, the man being sot-drunk and with a consistent history of in-shelter violence?
- What do you do with the guests who continue to sell drugs in-house?
- What do you do if your shelter does not admit unaccompanied minors, and Child Protective Services brings you a 16-year-old with borderline personality disorder who's been thrown out of *every* other emergency shelter in North Texas?

It's a real world out there and decisions have to be made . . . I am convinced that we can't do everything for everybody all the time. There have to be concessions; and such essentials as human rights, personal integrity, self-respect and other concepts have to be weighed with more mundane pragmatics and balanced in some workable fashion."

(Andrew Short, Executive Director, Texas Homeless Network, July 14, 1995)

Refuges for women

Men's shelters have a history as transient accommodation, functioning as a "charity to losers and failures who deserve only minimal support" (Single Displaced Persons Project 1983: 23). Women's refuges, on the other hand, developed more recently and are predicated on different assumptions. Initially, shelters for women were not necessary because women typically did not become absolutely homeless. They remained in the home, enduring abusive relationships, or they moved in with friends or relatives. With changes in divorce laws, relaxation of society's attitudes toward independent women, along with education about spouse abuse, it is now more common for women to enter shelters. Because women's refuges are a relatively recent phenomenon they are usually different from places for men – older institutions and some municipal shelters for women are glaring exceptions to these observations. Generally they are small (with fewer than thirty residents) and have a home-like atmosphere. Some shelters for men, on the other hand, have cots for hundreds of people in barracks-like settings. Women's shelters usually allow residents to come and go as they please. Length of stay rules are not rigid. Women may be permitted to stay as long as necessary to find housing or employment. Often founded on cooperative principles, women's refuges encourage residents to assist in cooking and maintenance and to participate in planning and management decisions. *Per diem* funding for women's shelters, which typically offer links with community and social services, is as much as three times higher than the subsidy for men's shelters that offer only basic accommodation. Staff-to-resident ratios are as low as 1:6. Women's refuges are frequently linked with transitional homes or permanent housing. As a consequence, residents are not forced into the destructive cycle of short shelter stays followed by life on the streets or tenuous short-term doubling-up arrangements.

A recent survey of women using hostels in Toronto found that two-thirds had stayed in a shelter previously, virtually all rely on social assistance, 62 per cent are single and only 8 per cent are currently married, 69 per cent indicated that they have been physically assaulted, and 53 per cent sexually assaulted; only 13 per cent acknowledged substance abuse. Just over half were born in Canada (Brown 1995: 58).

Women's refuges and transitional housing in North America have some of the most progressive and innovative programs designed to address homelessness. The experience of women in Britain is different because official attitudes toward women living independently have not kept pace with recent changes. Places for homeless women in England consist of three types: open-access hostels, refuges, and group homes. Open-access hostels generally accept all-comers unless they are drunk or blacklisted. Rents are customarily paid by government, but conditions are far from ideal. Watson and Austerberry found that the majority wanted to live independently, either alone or in shared accommodation.

> . . . [Many] were intensely miserable living communally, with little privacy or control over their day to day lives . . . To be homeless is to be unable to switch off the light when you want, cough when you want . . . It makes you feel terribly degraded . . . They had discounted renting or buying in the private sector where housing was in limited supply or beyond their means. Unless they were prepared to accept severely substandard, and often dangerous, housing offers, women had found that their chances of rehousing with the council or housing associations were negligible.
>
> (Watson and Austerberry 1983: 51)

The second type of shelter in Britain is women's aid refuges which are primarily for abused women. These tend to be small, communally run shelters, with democratic management. Funded by government, they are intended only for short stays as most women have young children and must be treated as priority homeless cases. Those without young children are left with few alternatives other than to make the rounds of shelters.

The third type of place provided by the voluntary sector for women consists of small hostels or group homes. These are usually privately-run establishments with a communal environment. Some group homes and residential communities provide permanent accommodation as an alternative to self-contained housing.

SUMMARY

The home should offer physical security and a source of stability. Having a home helps to define an individual's identity and status while providing a comfortable environment, essential to mental and physical well-being. The reality of B & Bs, welfare hotels, and emergency shelters departs dramatically from this ideal. Britain's B & Bs, used to shelter homeless families awaiting council housing, are crowded, noisy, unsafe, poorly maintained, and expensive. They lack privacy, security, and essential facilities.

The equivalent in the United States is the welfare hotel or motel. Used primarily in large cities, this alternative to the emergency shelter costs government ten times as much per household as does Aid to Families With Dependent Children. The negative effects of living in such quarters have been well documented. But they continue to be used, perhaps because welfare hotels are meant to discourage people from becoming "intentionally homeless" or trying to leapfrog over those waiting for scarce public housing units.

Emergency shelters or hostels, now common in all large North American cities, usually are large, barracks-like structures. Originally designed to remove homeless people from the streets, particularly during the coldest nights of mid-winter, they are now used year-round by families as well as single men, women and youths, many of whom have physical and mental health problems. The shelters, as a result, are chaotic, unhealthy, and dangerous. Though lacking the essential characteristics of homes they are being used as quasi-permanent residences because the range of perceived solutions is quite narrow and it appears that there are few other choices.

The discussion in Chapters 9 and 10 elaborates on these temporary expedients as responses to homelessness by governments and the voluntary sector are explored in Britain, the United States, and Canada.

PART III

RESPONSES BY GOVERNMENT AND THE VOLUNTARY SECTOR

Who should address the issue of homelessness? How have governments and voluntary agencies in Britain, the United States, and Canada responded to these problems? What are the results and implications? What are the advantages and disadvantages of the approaches taken in each country?

This section explores the evolving roles of national and local governments and the voluntary or third sector, and the tensions between the public sector's ambition to rein in social spending, the desire to devolve responsibility for some social programs to the voluntary sector, and the constraints placed on these third sector agencies by funding limitations.

Chapter 9 probes the relationships between central and local governments, the nature of public policies dealing with homelessness, gaps between policy and practice, and attempts to privatize the delivery of social services.

Chapter 10 examines the role of the voluntary sector which is typically responsible for local projects. Often, these are undertaken as "partnerships" with government, especially during the past decade, as the social welfare function has become privatized by public officials seeking to reduce or contain their welfare rolls and social budgets.

Chapter 11 examines the nature of innovative projects and programs initiated by government and the voluntary sector. Recommendations for precluding homelessness are found in Chapter 12. The final chapter offers concluding comments and comparative observations on responses to homelessness.

9

FROM SHELTERS TO PERMANENT HOUSING

The evolving role of government

> The doctrine of the minimal state, one restricted to such functions as promoting the conditions for a market economy to prosper, as well as maintaining law and order and defense, but returning many other responsibilities to individuals and lower-order social institutions is . . . central to the philosophies of many thinkers on the right.
>
> (Brenton 1985: 141)

This chapter presents an overview of the roles assumed by national and local governments in Britain, the United States, and Canada in addressing homelessness. The ways in which homelessness is defined and framed in public policy and discourse are analyzed, along with funding decisions and measures taken to implement housing and related social programs. Particular attention is given to the different responses by national and local states and by the voluntary sector – the role of this third sector is explored in greater depth in Chapter 10.

My research raises questions about the response of public agencies to homelessness:

- How is homelessness defined by public officials?
- What roles are assumed by governments in framing public policy, providing funding, and implementing housing and related social programs?
- What are the relationships between central and local governments with respect to homelessness?
- Has central government funding for local initiatives kept pace with the devolution of responsibility to local authorities?
- What gaps exist between policy and practice?
- How effective are public policies in addressing homelessness and related issues?

BRITAIN: GOVERNMENT'S RESPONSE TO HOMELESSNESS

Although homelessness was identified as a public concern in the National Assistance Act of 1948, government was reluctant to assume responsibility until substantive legislation was enacted in 1977. The 1977 legislation, as amended, charges authorities with the duty to prevent homelessness or to secure rehousing for people in the priority groups, provided that they do not have a "local connection" in another district and have not made themselves "intentionally homeless." Priority must be given to households with dependent children or pregnant women, and consideration should be given to persons who are vulnerable because of old age, physical or mental disability, or some other "special reason." For non-priority persons who are homeless or threatened with homelessness, the authority must offer "advice and appropriate assistance." Individuals with priority but without the necessary local connection must be put in contact with the appropriate local authority, and must be offered temporary housing if required.

The Housing (Homeless Persons) Act had an immediate impact. The total number accepted as being homeless in England and Wales was 182,000 *households* (with approximately 500,000 people) in 1992 (Central Statistical Office 1992a: Table 8.14). Half of those certified as homeless lost their homes. The other half never had a place of their own. During the 1980s about 1 million households (approximately 3 million people, half of them children) were registered as homeless. These "accepted cases" represented only half of those applying. One-third of those accepted were rehoused directly, another third were rehoused after a period in temporary accommodation, and the remaining third were placed in temporary units and then either dropped out or were rehoused via the normal waiting list (Department of Environment, H1 returns).

The principal public sector response to homelessness has been in the cities. Central government was a reluctant participant throughout the 1980s and early 1990s. Once national legislation was passed in 1977 the central issue was its implementation by a diverse array of local authorities. The key to this process is the *Code of Guidance*, issued by the Department of Environment under Section 12 of the Housing (Homeless Persons) Act of 1977.

Local authorities are required to "have regard" to the *Code* in implementing the statute; its provisions include several key definitions and obligations:

> A person is homeless . . . if he has no accommodation which he can occupy together with any other person who normally resides with him as a member of his family . . .

People are defined as homeless if they "are separated for no other reason than that they have no accommodation in which they can live together" or if they "cannot secure entry" to their dwelling, or "would be likely to be met with violence or threats of violence" if they tried to return to their dwelling.

> Authorities should not treat as intentionally homeless those who have been driven to leave their accommodation because conditions had degenerated to a point where they could not . . . reasonably be expected to remain – perhaps because of overcrowding or lack of basic amenities or severe emotional stress.
>
> Steps should be taken to ensure that children are not placed in care [i.e. foster homes] simply because the parents have become intentionally homeless.

> Permanent accommodations should be secured as soon as possible for those with priority need.
>
> Homeless people should not be obliged to spend a certain period in interim accommodations as a matter of policy.
>
> (Department of Environment 1977: 4.4)

While these policies represent a genuine commitment on the part of the government in 1977, there continues to be a considerable divergence between intent and practice. People housed under the Act, like pregnant or battered women, are visible and have high priority. Others, especially single people, generally are ignored despite being homeless or vulnerable. A shortage of council housing in some boroughs means that substantial numbers of families are placed for long periods in temporary accommodations. People are compelled to live in temporary accommodation (most of which is substandard or inappropriate in terms of location, size, or amenities) because there is a lack of affordable housing to rent in places where it is most needed.

Housing officials often assume that the problems of homeless and vulnerable people are solved once they are housed, regardless of the condition of the shelter provided. Donnison and Ungerson, though, found that "poor management and ill-considered allocation policies have created stigmatized 'sink' estates to which 'difficult tenants' are banished . . . As so often when dealing with people who lack political 'clout,' parliament has provided no appeals procedure" (Donnison and Ungerson 1982: 278–279).

A great deal of work remains to be undertaken at the national level in order to ensure that the legislation will be carried out fairly in all regions. Advocacy groups continue to urge a more active central government role in monitoring progress and consulting with councils regarding policies, practices and service standards. It is within this context that local government is charged with carrying out the intent of the Act.

The incidence of homelessness is higher in London than elsewhere in Britain. This is reflected in the organization, staffing, and resources of local authorities. Virtually all boroughs in the capital have specialist units looking after homeless households, while authorities elsewhere in the country combine homelessness and general administrative duties (Evans and Duncan 1988). When applicants represent themselves as homeless, agencies verify the information provided and conduct an investigation of their situation. The housing officer then must decide which unit of government is responsible, determine the extent of obligation, arrange for temporary accommodation or permanent shelter if necessary, and offer advice and referrals (Niner 1989: 16–22).

One of the most contentious clauses in the Housing (Homeless Persons) Act relates to intentionality, which renders applicants ineligible for aid. Interpretation of intentionality is left to local discretion. Half of the councils define those who move to seek employment as intentionally homeless; half do not. About 40 per cent deny accommodation to people who lack shelter as a result of rent arrears; the remaining 60 per cent accept them.

The Homeless Persons Act specifies that those in priority need must be housed; to qualify for housing a person must be both statutorily homeless and in a priority category. There are, however, considerable differences among authorities in their policies and procedures for complying with this statute. Persons found to be homeless and in priority need (as a per cent of total applications) vary from 26 to 84 per cent in different cities and boroughs (Niner 1989: 81). Disparities among these authorities cannot be explained by financial capacity; councils in poor districts are more likely to certify applicants as homeless. Better indicators are ideology or party affiliation. Labour-controlled councils are more generous than their Liberal or Conservative counterparts. London boroughs are more likely to certify as homeless unattached singles and childless applicants who are pregnant. Non-London authorities prefer households with dependent children. Other variables affecting certification decisions are housing shortages, the availability of temporary accommodation, vacancy rates, the levels of migration and immigration, and ethnicity. Three-quarters of all authorities certify as homeless people who apply from a women's refuge and half admit those from statutorily unfit dwellings; but only 30–40 per cent acknowledge people in overcrowded, bed and breakfast, hostel, or squatted accommodation. Urban officials are generally more liberal than their rural counterparts in making these decisions.

The law also encourages councils to shelter those considered vulnerable. There is no universal agreement on vulnerability, although the *Code of Guidance* indicates that an applicant may be vulnerable due to old age (60 years or older), mental illness or handicap, physical disability, or other "special reasons." Normally included in this discretionary category are battered women without children, frail people nearing age 60, and young people at risk of sexual or financial exploitation. In practice, many vulnerable individuals are not accepted as priority need cases, often because councils lack housing. Most local authorities – about 85 per cent – usually accept as homeless battered women without children or people leaving a long-stay hospital. But only 17 per cent usually accept an application from young single people on the basis of age alone (Niner 1989: 81–82).

Part II of the 1985 Housing Act (an amendment to the 1977 legislation) instructs local authorities to give "reasonable preference" to homeless persons for council units. People who are desperate complain that they have only been offered housing which other families have refused. The types of temporary and permanent accommodation used to shelter homeless people run the gamut from caravans (trailers) to permanent council dwellings.

Part III of the 1985 Act specifies that councils have a responsibility to those who are "threatened with homelessness within 28 days." The great majority of vulnerable individuals are dealt with before the 28 days expire. Preclusive actions include welfare benefit and tenancy advice, liaison with landlords and building societies (i.e. lenders), and mediation. Birmingham has a squad of homelessness officers who deal with harassment and tenancy issues. Other cities offer financial and budgeting advice to clients facing eviction because of rent arrears (Niner 1989: 41).

Government cutbacks in housing expenditures during the 1980s eroded local capacity to provide and manage dwellings. Housing's share of public

A lack of affordable housing

The overriding cause (of homelessness) is that people are unable to find or retain housing at rents or prices they can afford and with security of tenure . . . it is still to rented housing that homeless individuals and families are compelled to look for a solution to their housing problems. In practice this means local authorities and housing associations.

(Greve with Currie 1990: 1–2, 16)

expenditure fell from 7 per cent in 1979/80 to less than 3 per cent in the late 1980s. Faced with falling subsidies and rising costs, councils were forced to raise rents or to make transfers from the general rate fund to their Housing Revenue Accounts (Central Statistical Office 1990: Tables 3.6, 3.10).

To cope with shortfalls local councils used temporary accommodations for homeless households – 80 per cent of them in London and the Southeast. In February 1992 the government established an Empty Homes Agency to fill vacant properties with homeless households; at the time there were 768,000 empty properties in England and Wales, half of which were thought to be suitable for immediate occupation (Audit Commission 1989: para. 22). By the early 1990s a trend away from hostels and toward short-life housing, under lease for up to three years, was perceptible. Meanwhile, London boroughs established standards for bed and breakfast establishments to ensure better conditions in temporary accommodation. In addition, the percentage of new rentals let to homeless people in the early 1990s represented an increase over the 1980s (Department of Environment 1991).

In January 1994 the government issued a Green (consultation) Paper, known as the Homelessness Review. The key principles of this Review, incorporated in the Housing White Paper of June 27, 1995, are:

- Local authorities will no longer be permitted to house homeless households solely because they are homeless.
- Local authorities' duties to provide emergency assistance to homeless people will be weakened.
- Council housing will be allocated from a single waiting list on which homeless people will have no particular priority except on the basis of time spent on the waiting list.
- Social housing will be privatized where possible, with stock transferred from local authorities to housing associations and private entities.
- The owner-occupied sector will be enlarged; tenants of charitable housing associations will be given the right to buy.
- Restrictions on private landlords' right to evict tenants will be eased.
- Tax concessions will be made to institutional investors in private rental housing.
- Funding for the Rough Sleeper's Initiative will be extended to include areas in addition to London.

(Department of Environment 1994b; 1995)

Critics have expressed serious misgivings about these new policies because an underlying cause of homelessness – the lack of affordable rental units – is not addressed. (For example, a 1994 Shelter survey found that local authorities place homeless households in private rentals where the average weekly rent is £105, compared with average weekly council rents of £33 and average housing association rents of £45 (Shelter 1994: 13).) Many are fearful that local authorities will turn their backs on homeless people. Officials now have the discretion to refuse to accommodate demonstrably homeless households when there are private rental units available in the district. Moreover, because local authorities no longer have to provide permanent housing it is expected that some households will face repeated homelessness and frequent moves, which are disruptive for children in school and for wage-earners who use public transit between home and work. The government has been criticized as well for extending home ownership while simultaneously reducing mortgage assistance to low-income home-owners. Some expect that repossessions of owner-occupied dwellings will rise, placing more pressure on social rental housing (Shelter 1995: 1–4).

THE RESPONSE BY BRITISH LOCAL AUTHORITIES: CASE STUDIES FROM LONDON

London, with a population of 7 million, is a global city, the seat of government, the headquarters for major companies and financial institutions, and a center for tourism. But it has changed dramatically in recent years. There are twice as many people working in financial services as in manufacturing, once a dominant activity. The city is sharply split along lines of income, class, race, age, and ethnicity. There is a gaping divide between well-paid workers in higher socio-economic groups with relatively secure jobs and those in tenuous positions with low pay and benefits. About one-third of the city's households are singles, many of them poor, elderly, or members of minority groups. As they have become marginalized by structural changes in the labor market, they have found that their access to suitable, adequate, and affordable housing is restricted.

Land values in the city have risen sharply, partly a result of gentrification, thus making the provision of low-cost housing all but impossible in central London where, paradoxically, homelessness and poverty are concentrated. Some of the country's worst housing estates are located in close proximity to the new luxury developments. At the beginning of the 1990s unemployment rates were above 14 per cent of the economically active population in poor parts of the city, notably Hackney and Tower Hamlets.

In 1989 local authorities in Greater London accepted as homeless a total of 33,170 households, representing a rate of 12.4 per thousand population, an increase of 38 per cent (1983–1989). In the city's central boroughs the rate of homelessness was 25 per thousand.[1] Half of those accepted as homeless are from ethnic minorities (*Housing Association Weekly*, February 9, 1994). Estimates of the numbers of single homeless people in London range from 64,500 to 78,000; these individuals are sleeping rough, squatting, or living in precarious temporary accommodations. In addition, as many as 250,000 young singles, women, and ethnic minorities who rely on public benefits or low-wage jobs are thought to be involuntarily doubled up because they do not qualify for council housing and cannot afford private rentals.

Affordability has become a central issue for a growing number of London households. House prices almost doubled from 1985 to 1990, an increase of 87 per cent; council rents were raised by 66 per cent, and private rents increased even more rapidly, following the deregulation of the private rented sector in 1988. The amount of housing benefit required to cover private rents rose by an average of more than 120 per cent between 1988 and 1993; it is now 75 per cent higher than the subsidy to a council tenant. As a result, council housing is becoming residualized, accommodating mostly those who are receiving public benefits (Maclennan *et al.* 1990). These trends are likely to continue as a result of the 1989 Local Government and Housing Act which linked rents to house prices and attempted to make council housing self-financing.

During the 1980s and early 1990s councils faced increasing demands on their services at the same time as their scope for action was circumscribed by the government. Changes affecting their responsibility were intended not only to control local borrowing power, but also to curb spending. National housing policy has discouraged new construction. Although more than 150,000 council dwellings were sold in London during the 1980s, the use of revenues from sales was restricted, and local authority completions fell from 15,000 in 1980 to 1,000 in 1990. Simultaneously, there was a reduction of 84 per cent (in real terms) in Housing Investment Programme allocations from central government to London boroughs, and a 44 per cent decline in subsidies to finance public sector investment. Nick Raynsford, MP for Greenwich, illustrated the scope of the problem in 1993: local authorities in London built sixty-eight new homes, housing associations started 4,800, and 34,000 households were accepted as homeless.

Remedies for this situation have been suggested by the Association of London Authorities and various research organizations. They include the provision of additional council housing stock by rehabilitation and new construction (where land is available), helping owners with mortgage arrears, and restoring security for private tenants. Resources would be directed at increasing housing inventories (while creating jobs for residents), rather than on the costly short-term approach of leasing temporary accommodations for homeless households. Suggestions have been made to allow local authorities to use public land for non-profit housing, especially in inner urban areas, to monitor the operation of private landlords and other housing providers, and to increase accountability, user control, and participation in the planning and management of housing.

The residualization of council housing and the nature of the difficulties confronted by large urban authorities are illustrated by three cases from the London boroughs of Haringey, Westminster, and Tower Hamlets. Haringey received an increasing number of priority need applicants in the 1980s but was unable to meet demand. The borough acquired units, provided loans for others, converted dwellings to rentals, and made renovation grants. Its net additions to stock in recent years, however, represent only about 2 per cent of the more than 12,000 households on the waiting list. Like three-quarters of London's boroughs Haringey has found it necessary to look elsewhere in the city for additional housing.

Westminster tried to discourage new applicants by placing priority need households far from the city and by seeking exemption from its obligations under the Homeless Persons Act. In 1991 the borough offered councils in the North of England £400 a year for every homeless family accommodated. Later it froze the transfer list – 1,700 families were still on the waiting list in 1993 – and removed 300 households previously found to be in priority need. There were more than 1,335 families in temporary accommodation in 1993; 80 per cent were sheltered outside the borough. Its practice of selling council flats, called the "Building a Stable Community" policy, included boarding up half the council flats, as they became vacant, until sold to owner-occupants. Some were fitted with steel doors to deter squatters and the council ripped out bathrooms to discourage occupation.

Westminster's Housing Committee Chairman defends this strategy as a means of achieving better housing for less money. The council hopes to reduce both its housing and social service costs by placing homeless applicants at such a distance that it is impractical for them to return when they need help. The borough could use its legal powers to obtain "planning gain" funds for affordable housing from new projects in Westminster, but has declined to do so.

Tower Hamlets is the major destination for Bangladeshis in Britain. By virtue of the government's policy of devolving responsibility to the local level, the borough has an enormous financial burden, coupled with a very limited land supply and high land costs. Over 80 per cent – 50,000 units – of the housing stock in the borough is council-owned. It has the highest number of homeless families living in B & Bs and the greatest amount of overcrowding in Britain (10 per cent for all families; 67 per cent for immigrant families). Bangladeshis comprise 10 per cent of the borough's population but 69 per cent of the area's homeless households. Though they live in the poorest housing and have the highest rates of homelessness, Bangladeshis represent only one-fourth of the waiting list for council housing. Instead they have been sent to B & Bs far from London.

The Commission for Racial Equality investigated to determine why only 18 per cent of Bangladeshis (compared to 43 per cent of whites) were rehoused permanently within three months after being declared homeless. They discovered that most of the Bangladeshi families had been put up in single rooms far outside London at Southend-on-Sea, where the former head of the Tower Hamlets Homeless Families Unit had established a hotel business. The Commission found that some rooms, for which the borough paid five times the normal room rate, lacked heating, cooking facilities, fire alarms, and curtains. Only one in nine properties had a telephone. Tower Hamlets was issued a discrimination notice and acknowledged "with hindsight that the policy of sending homeless families to Southend was a disaster and that it was even less appropriate to put Bangladeshi families in Southend" (Commission for Racial Equality 1984).

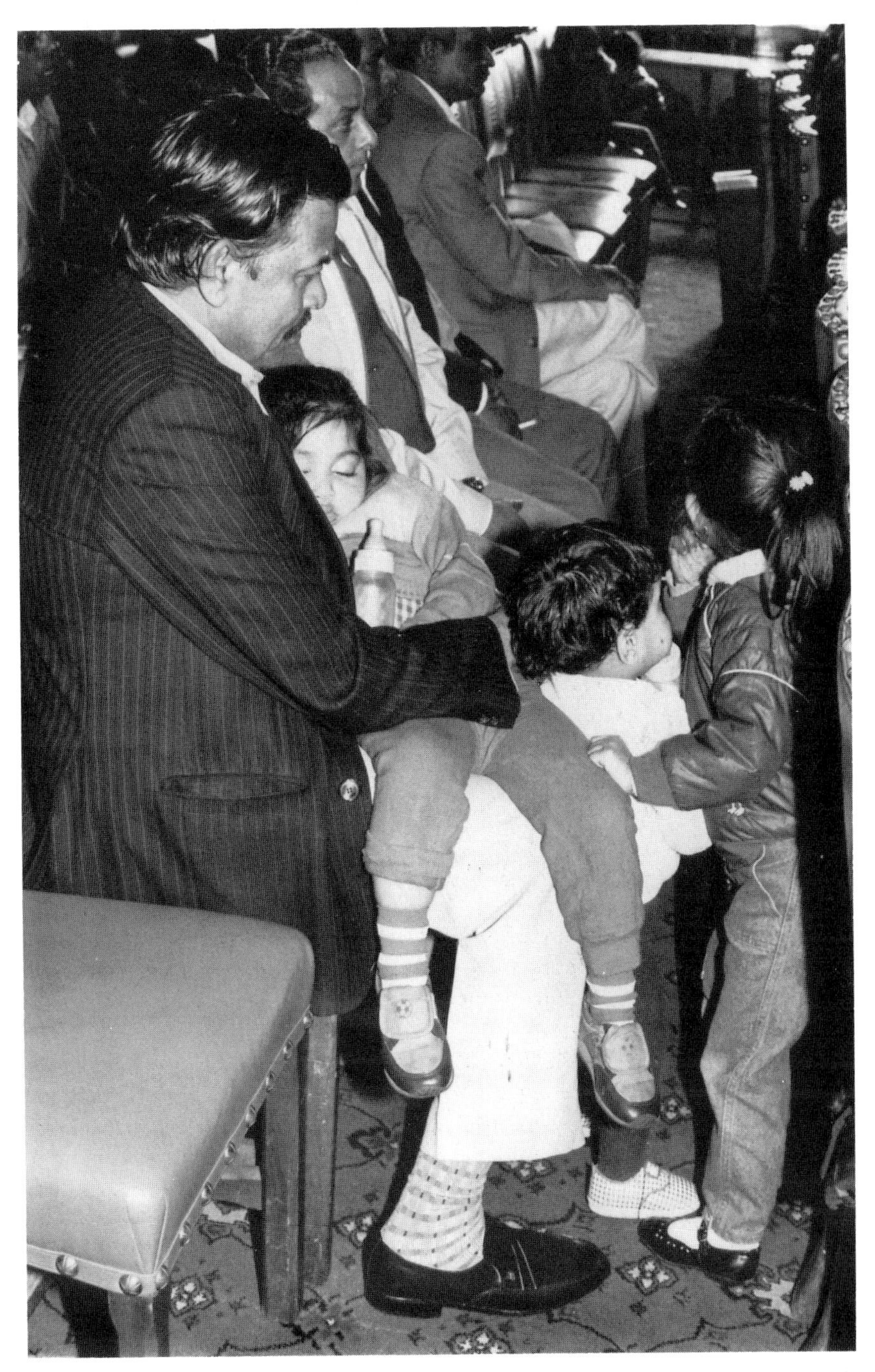

38 Homeless Bangladeshi families wait for housing in the London Borough of Tower Hamlets (Photograph by Jonathan Stearn, courtesy of Shelter)

Subsequently (1989) Tower Hamlets adopted a "sons and daughters letting policy" which gave priority in housing allocation to the children of existing council tenants, mostly whites. The Commission for Racial Equality secured a court order against Tower Hamlets Borough Council in 1991 in order to carry out the requirements of the non-discrimination notice served on the Council four years earlier. In April, 1992 the Court ruled that Tower Hamlets was not entitled to refuse applicants on grounds of their immigration status (Commission for Racial Equality 1993).

THE UNITED STATES: GOVERNMENT'S RESPONSE TO HOMELESSNESS

American housing policy in the 1980s consisted of stopgap measures and experiments with privatization and supply-side devices. Both public and voluntary sectors concentrated on such expedients as emergency shelters and welfare hotels. The federal government significantly reduced expenditures on housing (by more than 70 per cent), employment and training programs (by 70 per cent), block grants for child care (30 per cent), and grants to state and local governments, by one-third (Wolch and Akita 1989).

The Reagan administration's initial reaction to the growing numbers of homeless people was to assume that voluntary organizations would look after them. The lack of formal response mirrored the administration position that social programs should be reduced; one-third of the spending cuts made in Washington were in social spending although these areas represent less than one-tenth of all federal expenditures. By 1983 when it became apparent that homelessness required national attention the Secretary of Health and Human Services established a coordinating body, the Federal Task Force on the Homeless, and Congress appropriated emergency funds (through the Federal Emergency Management Agency) for food and shelter on a temporary basis. The Reagan Administration, however, declined to seek funding to extend this initiative. In 1984 HUD issued its first report on shelters and homelessness (U.S. Department of Housing and Urban Development 1984). Shortly thereafter, the Department of Defense was authorized to use certain military facilities as emergency shelters (U.S. General Accounting Office 1985).

It was not until 1987 that comprehensive legislation was enacted. The Stewart B. McKinney Homeless Assistance Act (Public Law 100–77, as amended by Public Law 100–628 in November 1988) resulted from pressure by advocacy groups and Congressional supporters. This legislation represented a dramatic reversal by Reagan; Title I of the Act acknowledges that:

> The nation faces an immediate and unprecedented crisis due to the lack of shelter for a growing number of individuals and families . . . the problem of homelessness has become more severe and, in the absence of more effective efforts, is expected to become dramatically worse . . . there is no single simple solution to the problem of homelessness because of the different sub-populations of the homeless . . . the causes of homelessness are many and complex . . . States, units of local government,

> and private voluntary organizations have been unable to meet the basic human needs of all the homeless and, in the absence of greater federal assistance, will be unable to. The federal government has a clear responsibility and an existing capacity to fulfil a more effective and responsible role to meet the basic human needs and to engender respect for the human dignity of the homeless.
>
> (U.S. Congress 1987: Sec. 102).

Title II of the McKinney Act established an Interagency Council on the Homeless within the Executive Branch to coordinate the activities of federal agencies responsible for a variety of housing assistance programs. These include grants for emergency shelters, transitional housing, permanent dwellings for handicapped homeless people, and Section 8 (rental voucher) funding for new single room occupancy units. Programs under the Act are administered by nine federal agencies, among them the Departments of Housing and Urban Development, Health and Human Services, Labor, Education, and Veterans' Affairs, the Federal Emergency Management Agency, and the General Services Administration.

The framers of the McKinney legislation recognized that they must deal with much more than housing: originally the Act provided funding for twenty programs, including community mental health services, job training and education, mental health and substance abuse demonstration programs, emergency food assistance, community action, and veterans' assistance. On the basis of local assessments of need municipalities are expected to formulate appropriate solutions for their own communities. In order to be eligible for federal grants each participating jurisdiction is required to create a Comprehensive Homeless Assistance Plan and, for some grants, matching state (or state and local) funds.

This comprehensive approach led some observers to characterize the McKinney Act as "a complex piece of social legislation" which signaled the federal government's acceptance of "the view that the challenge of the growing crisis requires additional help from the federal government." Some speculated that "the Act's stance towards homelessness and its specific substantive and procedural remedies potentially set a precedent for the future role of the federal government in the issue" (Keyes 1988: 1, 13). Subsequent appropriations were much lower than expected, however, and the Reagan administration's commitment in practice was lukewarm.

Wolch and Akita concluded that the legislation "does not address the economic and social-welfare-policy roots of the homelessness crisis." The government's response was criticized because it did not replace severe budget cuts in social programs, nor did it include mandated programs. Moreover, the McKinney "programs were mostly targeted toward emergency needs rather than long-term solutions," and they "seriously underestimated the real resource costs of solving the crisis in an efficient and humane way" (Wolch and Akita 1989: 62–85).

President Reagan's policies were not substantially changed by his successor. When George Bush assumed office in January 1989, his agenda was blunted by a burgeoning federal deficit, a series of foreign crises, the $500 billion savings and loan debacle, scandals in HUD, and the perceived need to "hold the line on taxes." His funding requests in pursuit of the McKinney legislative mandates were fainthearted, reflecting an unfocused and tenuous urban program. Moreover, in 1992 Bush vetoed the urban aid tax bill, intended to stimulate the production of affordable housing.[2]

Several low-income housing initiatives were undertaken during the early 1990s; but implementation was timid.[3] The principal federal programs addressing homelessness in the early 1990s were for emergency shelters (including a provision allowing use of up to 20 per cent of the funds for services and prevention activities), transitional housing, Section 8 vouchers for single room occupancy units (SRO), and "Shelter Plus Care," which links housing with social services, particularly for people who are mentally ill or substance abusers. The Interagency Council which coordinated the efforts of federal agencies concerned with homelessness lost its appropriation in 1994. Appropriations for fiscal years 1993 and 1994 are shown in Table 9.

In addition to these HUD programs, there are a number of federal initiatives which serve segments of the homeless population. These include measures to deal with rural homelessness, AFDC Emergency Assistance programs, and funding designed to assist persons with HIV-AIDS. Alcohol, drug abuse and mental health programs for the poor may also be used to aid homeless individuals. Some federal agencies, like the Veterans' Administration, have increased their efforts to address the needs of homeless clients, but funding has not been commensurate with need. In late 1993, for instance, the VA announced a new $5 million initiative – a paltry sum to deal with mentally ill veterans across the country.

In late 1993 the Secretary of HUD under President Clinton, Henry Cisneros, outlined the Department's commitments for the coming fiscal year. This change in direction was based on a critical appraisal of past efforts:

> Current assistance to homeless families and individuals is fragmented . . . There is often no unified system which provides services for the different groups of the homeless population. There is limited effort to balance the number of transitional units and permanent units available for homeless persons. A community's homeless system must create needs-specific facilities, housing and services for families and for individuals. This strategy should be designed locally and federal programs must, in turn, support the design of this approach.
>
> The competitive HUD homeless programs, which represent the majority of Federal targeted homeless assistance funds, complicate the fragmentation of homeless assistance by having different requirements and purposes. As a result, providers often must design programs to meet funding requirements rather than actual community needs. Funding should be responsive to actual need.
>
> (Low Income Housing Information Service 1993: 3)

The thrust of HUD's initiative is to implement the "continuum-of-care" model which had been introduced in Washington, D.C. under the DC/HUD

Table 9 Federal appropriations for HUD homelessness programs (in $ millions)

	FY 1993 actual	*FY 1994 appropriation*
Emergency shelter	50	115
Transitional/perm. housing	150	334
Section 8 – SRO	105	150
Shelter plus care	267	124
Total	572	723

Source: U.S. Department of Housing and Urban Development (1995: 3)

Homeless Initiative. This concept "brings the homeless into a system, assesses needs and provides those persons with a full range of services, if any, needed to regain independent living." HUD's proposal includes establishment of an Innovative Homeless Fund in large cities with "supplemental funding for continuum-of-care homeless assistance," consolidation of the five separate McKinney HUD Homeless Assistance programs, and development of a federal strategy on homelessness (Low Income Housing Information Service 1993: 3).

Slightly over a year later, in January 1995, in reporting to Congress on the McKinney programs, HUD found evidence of progress: "Most emergency shelters have gone from shelter only to include supportive services" (U.S. Department of Housing and Urban Development 1995: iii). Approximately 7 per cent of assisted households were able to find permanent housing (at a cost of $45 per person per day). A diverse array of support services were offered by grantees, 80 per cent of them by non-profit organizations "because they were unencumbered by a government bureaucracy and could draw on a wide variety of other community resources" (U.S. Department of Housing and Urban Development 1995: iv). The report to Congress indicated that programs are successful when they focus on the individual's specific needs, respect individual rights, require individual responsibility, and are comprehensive. HUD concluded, however, that

> the sheer number and variety of HUD's McKinney Act programs have sometimes created barriers to their efficient use. Differences among the target populations, eligible activities, application requirements, and selection criteria for the various programs have made this assistance difficult to obtain and coordinate. Overlapping sets of regulations and reporting requirements have further complicated grant administration. Moreover, the unpredictability of competitive grants, appropriation levels, and the varying lengths of grant awards have frustrated attempts to implement longer term, comprehensive strategies for eliminating homelessness.

In one case "as few as 3 per cent of applicants (42 in 1,400) were funded because of large needs and limited appropriations" (U.S. Department of Housing and Urban Development 1995: iii).

With the election of right-wing Republicans in 1994, however, HUD's future became uncertain. Newt Gingrich's "Contract with America" was intended to severely cut housing and social programs. In late 1994 House

Speaker-elect Gingrich made his intentions clear: "The weak political constituency for the Department of Housing and Urban Development makes it a prime candidate for cuts . . . " (*Washington Post* December 13, 1994: 1).

THE RESPONSE BY STATE AND LOCAL GOVERNMENTS IN THE UNITED STATES

The political system in the United States is considerably more complex than in either Britain or Canada. There are more tiers of government and quasi-governmental organizations, a variety of different public management models, and a wide diversity of cities and regions in terms of history, location, climate, ethnicity, and economic well-being. Center–periphery disparities are more apparent than in Canada or Britain. Striking differences exist between inner cities and suburbs in terms of housing, income and wealth, employment, and family structure. The issue of homelessness is generally perceived as an inner city problem. Not surprisingly, then, substantial variation is evident in the responses of local government to homelessness. Some state and municipal agencies choose to deal with homeless persons by adopting a comprehensive assortment of innovative programs. Others ignore the issue. Most states, though not directly involved in providing services to this population, pass through federal grants to local governments.

Social programs for poor people in the United States generally are controlled to some degree by state and local governments. A number of assistance funds are optional and states may opt out. As the federal deficit ballooned in the 1980s support for state and local governments was cut. Because most states are required to balance their budgets, they were compelled to reduce spending, raise taxes, or both, to compensate for reduced federal spending. Cash assistance programs include Aid to Families with Dependent Children (AFDC), the Supplemental Security Income Program (SSI), and general assistance (welfare).[4] AFDC illustrates the problems which occur when states are allowed free rein in setting social policy. AFDC is jointly funded by federal and state governments; the states pay an average of 46 per cent of program costs. Median benefits in 1991 for a family of three with no other income were $367 monthly – about 42 per cent of the average poverty line. But the benefits ranged from a low of $120 per month in Mississippi to a high of $924 per month in Alaska. As a result of these huge differentials poor families sometimes move to another state in order to obtain higher payments.

Over time AFDC benefits are declining – a drop of 42 per cent in real terms between 1970 and 1991 (Ellwood 1993: 4). States are also cutting back on payments which help homeless families or those who are extremely vulnerable. The AFDC Emergency Assistance Program (which receives matching federal funds) provides money to preclude eviction, to prevent utility shut-offs, or to help with rental deposits. In 1991, ten of the thirty-two states participating in this program reduced their commitments, while five states established new or slightly expanded schemes (Center on Budget and Policy Priorities 1991).

Twenty-nine states have emergency housing programs for homeless people. Affordable housing programs in twenty-six states offer rental assistance and

A San Francisco mayoral candidate speaks out on vagrants

"The problem of the vagrants, panhandlers and bums who plague our neighborhoods and commercial districts can be controlled, if there is the political will to do it. Homeless advocates would have us believe that the homeless are just like you and me, except that they lack a roof over their head. This ludicrous assertion is simply NOT TRUE!! . . . Vagrants are vagrants because they either don't know how, or can't, or choose not to, live like the rest of us.

San Francisco has become a magnet for vagrants from all over the country. Our weather is comparatively good for living on the street, our general assistance payments are high, and most of our politicians have a welcoming, politically-correct attitude of tolerance. As a result, we take vagrancy on our streets for granted. It isn't until we travel to another city that we recognize the problem for what it is. Other cities have addressed the problem and controlled it. Seattle's problem decreased when generous cash payments were lessened and replaced with mandatory assignments to drug and alcohol detox centers. In Washington, D.C., police admonish and may cite citizens who give money to panhandlers . . . The laws of supply and demand have infested downtown, our neighborhoods, and our parks with an ever-growing population which threatens to eventually bankrupt our city and destroy our quality of life . . . We must re-examine how we're spending our money. The number of vagrants on the street, estimated to be from 6,000 to 10,000, is not decreasing, but increasing.

We have travelled in the wrong direction for far too long. We must face the reality that vagrancy is against the law, and proceed accordingly with formulating public policy . . . People do not have the right to live on our streets or in our parks. Proliferation of vagrancy is a problem of law enforcement, and should be dealt with accordingly . . . The free ride is over in San Francisco."

(Ben L. Hom, September 1995)

encourage rehabilitation and construction of affordable units. Another sixteen states have housing trust funds for new construction. A number of states made significant cuts in these programs during the early 1990s because their revenue sources dwindled while welfare caseloads expanded dramatically. This was particularly true of industrial areas in the Rustbelt. Michigan abolished special needs payments to AFDC mothers and eliminated its general assistance program and its medical assistance scheme, leaving 82,000 people – including about 10,000 with disabilities – without cash assistance or health care coverage.

Cities in the United States are confronted with myriad problems: racial divisions, the flight of middle-class blacks and whites to the suburbs, deteriorating municipal services and failing public school systems, violent crime, drugs and poverty.[5] As creatures of the states, municipal governments have budget and tax limitations which constrain their ability to spend. Enormous disparities in social expenditures are the result. In 1991, for example, per capita spending on housing and community development was more than thirty times higher in Washington, D.C. ($402.63) than in Dallas ($12.86) and almost ten times higher than in Chicago ($42.66) (J. Kifner, *New York Times*, February 12, 1995: 30).

Local governments deal with homelessness by operating public shelters, funding private or voluntary groups to provide hostels and other services, distributing vouchers to homeless individuals for use in motels, hotels, or apartments, and leasing or rehabilitating buildings for private shelter operators. The principal municipal initiative with respect to homelessness has been to provide shelters or to fund third sector shelter providers. A 1989 HUD report reviewed the efforts of the nation's five largest cities: New York City, Los Angeles, Chicago, Philadelphia, and Houston (U.S. Department of Housing and Urban Development August, 1989). It revealed that there are significant differences among municipalities in terms of the incidence of homelessness, funding, shelter capacity, and the ability to serve homeless people. Two of the five (New York and Philadelphia) are "shelter entitlement" cities where all citizens have an unrestricted right to emergency shelter (as a result of court orders); in a third municipality, Los Angeles, entitlement is restricted to those who apply for welfare assistance.

Comprehensive Homeless Assistance Plans (CHAPs) provide information about the homeless population and how McKinney Act programs can be used to meet local needs. The estimates of homelessness from the CHAPs are:

New York City	35,000–90,000 persons
Los Angeles	35,000
Philadelphia	12,500
Chicago	12,000–25,000
Houston	3,000–15,000

In the early 1980s federal aid was concentrated on emergency shelter grants and food. In recent years, however, funding has been redirected to services related to health care, job training, and psychiatric counseling. In Chicago, about three-quarters of public spending on homelessness is concentrated on services; in Philadelphia the comparable figure is over 60 per cent.

The total amount of public expenditures (in millions of dollars) on homeless people in the five cities is shown in Table 10 for 1988 by local, state, and federal governments.

Traditional shelters for singles are usually located in Skid Row. Newer, smaller facilities are situated in outlying areas. As a result, the majority of beds are in central locations near downtown, but newer shelters and transitional housing are dispersed. In Chicago, for instance, in 1984, 90 per cent of the shelter beds were downtown. Since then, two-thirds of new beds have been situated in outlying areas. In New York City, one-third of shelters and one-half of total shelter beds are in Manhattan. Like other American municipalities the City of New York has encountered virulent resistance from residents of suburban neighborhoods.

The trend in the United States is toward smaller, more specialized shelters; but small places still represent a minority of all shelter beds. Philadelphia's system consists of small shelters, most of which are personal care homes run by family businesses. Most of Chicago's shelters are relatively small; but about half of the city's shelter beds are in institutions with more than a hundred beds. The numbers of homeless people in New York City are so high that large shelters are the norm. While two-thirds of the shelters in Los Angeles are small, 80 per cent of the city's shelter beds are in large facilities. In general,

Table 10 Public expenditures on programs dealing with homelessness, 1988

	Local	*CDBG*	*State*	*Federal*	*Total*
New York City	375	40	230	113	758
Los Angeles	13	1	14	16	44
Philadelphia	23	0.2	12	8	44
Chicago	4	0.3	2	10	16
Houston	0	0.5	0	5	5

Source: U.S. Department of Housing and Urban Development (August 1989)
Note: The column labelled CDBG denotes Community Development Block Grants from the federal government; the column labelled Federal denotes federal expenditures exclusive of CDBG.

private, transitional, and specialized shelters are small, while public emergency shelters are large. The number and allocation of shelter beds in each city is shown in Table 11.

Most shelters subsist on a combination of public funds and grants from foundations or charities. In a few cities the municipality runs shelters as well; the City of New York operates 10 per cent of its shelters, representing one-third of all shelter beds. Of the shelter beds in New York City, 90 per cent receive the majority of their funds from the public sector. Most shelters in Los Angeles (85 per cent) and Philadelphia (95 per cent) get a portion of their funding from the city treasury. But only 60 per cent of the shelter beds in Houston have access to public funding and these contributions cover less than one-half of expenses. Houston relies heavily on the private and voluntary sectors (especially the United Way) for support.

There are two types of shelters: emergency and long-term. The latter category consists of transitional and specialized treatment facilities, which have three characteristics: they allow extended stays; they provide a broad array of services; and they focus on enhancing residents' self-sufficiency. Transitional facilities tend to concentrate on housing and employment while specialized services focus on rehabilitation of those suffering from mental illness or substance abuse.

In all five cities most of the beds are in emergency shelters: this proportion ranges from 60 per cent in Chicago to 90 per cent in Los Angeles. This pattern reflects the fact that, until recently, the facilities for homeless people offered few services and served only short-term, emergency needs.

Table 11 Number and type of shelter beds in five U.S. cities, 1988

	Shelter beds	*For families (%)*	*For individuals (%)*
New York City	30,500	62	38
Los Angeles	10,332	30	70
Philadelphia	6,936	34	66
Chicago	2,588	37	63
Houston	3,168	29	71

Source: U.S. Department of Housing and Urban Development (August 1989)

At least 30 per cent of shelter beds in the cities surveyed are intended for families. Despite increases in services for families during the 1980s, they remain under-served; the demand for family shelters is outpacing that for singles' facilities. This trend exists, in part, because many shelters will not admit children under age 12. The majority of shelter beds are restricted exclusively to men or to women. The number solely for men ranges from 77 per cent in Houston and Chicago to 96 per cent in Los Angeles. During the late 1980s, however, most cities began to provide more spaces for women, particularly for those who have been battered. Spaces for people with substance abuse problems and those with multiple disabilities (such as mental illness, alcoholism, and AIDS) are inadequate to meet demand.

In "shelter entitlement" cities homeless people are allowed to stay as long as they require a bed. In other cases the length of stay varies dramatically from place to place, ranging from one night to eighteen months. The average stay is thirty-four days in Los Angeles and twenty-three days in Chicago. In New York City 20 per cent of the families remain in the system for over eighteen months, and 4 per cent for more than three years. In Philadelphia the average stay is nine months for families and six months for individuals. It is difficult to determine the amount of time spent in shelters by people who move from one facility to another. This "revolving door" syndrome is common, especially in cities which lack shelter entitlement legislation.

While most cities concentrated on such expedients as shelters and food banks, a number of local government agencies created innovative projects and programs. Though some of these efforts do not deal directly with homeless people, they relieve shelter problems by increasing the stock of affordable housing and may help to preclude homelessness.

In Boston, San Francisco, Seattle, Washington, and other cities, a variety of programs are used to harness the potential of large downtown commercial projects. In return for planning approvals, developers of these complexes contribute to the construction of low-income housing, day care, and related facilities in poor neighborhoods.

In a few cities, notably Philadelphia, Los Angeles, and Portland, Oregon, joint public–private ventures create SRO units, refuges or transitional housing. Pittsburgh built SRO housing in a program involving state and local governments, a private corporation, and the use of federal tax credits for low-income housing and historic rehabilitation.

Cook County and the City of Chicago developed a program to produce a substantial number of renovated dwellings. The county transfers foreclosed buildings to the Chicago Housing Partnership which induces businesses to invest equity capital in return for tax incentives. The City of Chicago offers deferred-payment second mortgages and rent subsidies for residents who move into the rehabilitated houses.

In many communities group homes, hostels, and shelters are voted down by residents who feel that their property values and the safety of their families will be threatened by the influx of homeless people. In a few instances, however, voters agreed to increase taxes to finance low-income housing. Dade County, Florida imposed a commercial real estate transfer tax to pay for low-income housing. Seattle voters passed a housing levy which raised property taxes in order to finance $50 million of new housing units over a period of eight years.

CANADA: GOVERNMENT'S RESPONSE TO HOMELESSNESS

Until 1986 the Canadian federal government's role in dealing with homelessness was passive. Starting in 1984 an effort was made to devolve responsibility for housing and related issues to the provincial and municipal levels. The national housing agency, Canada Mortgage and Housing Corporation (CMHC), experienced budget cuts and a narrowing of its focus, though it continued to assume part of the burden for funding social housing. Agreements were signed which allow the provinces and territories to run non-profit housing programs. If a non-profit group's proposal is accepted by the provincial government, federal officials are committed to sharing the difference between the project's operating costs and the amount paid by tenants. Results, though, have been disappointing. In 1989 CMHC announced reductions in cooperative and rental housing programs in order to deal with a mounting national debt. Moreover, with the introduction of severe budget cuts in 1995, the future of existing social housing programs is in doubt.

In Canada the attention given to social housing and homelessness varies profoundly from one region to another. Some provinces have taken little action. Only one, British Columbia, has adopted legislation, the 1994 Homeless and At-Risk Program. Others, like Ontario (under the Liberals and then the New Democrats, until 1995), demonstrated that a great deal can be accomplished when all three levels of government cooperate. Starting in the mid-1980s Ontario introduced several new measures. It passed the Rental Housing Protection Act to control demolition and conversion of rental housing when the supply of affordable units is threatened. It conducted an inquiry into the housing situation of roomers, boarders and lodgers which resulted in the inclusion of rooming houses and SRO units under the Landlord and Tenant Act. A housing policy statement was devised, requiring all municipalities to allocate at least 25 per cent of new development and redevelopment for affordable housing.[6]

The provincial government's main thrust, prior to the election of a Conservative government in 1995, was to produce permanent housing for low-income households by providing grants to municipal non-profit and private non-profit housing corporations, as well as to housing cooperatives. Approximately 80 per cent of these units are subsidized. Roughly half of the subsidized dwellings are allocated to very low-income households while the remaining half go to moderate-income households who receive a "shallow" subsidy. Within a particular building or project developed under these guidelines the mix of units is 35–40 per cent for "deep core" poor households requiring a substantial or full subsidy, about 30–40 per cent for "shallow core" households receiving a smaller subsidy, and the remainder for people paying full market rents. There is no differentiation between subsidized and non-subsidized dwellings. In places like Toronto's St. Lawrence neighborhood, these policies have allowed the creation of new communities in the inner city with a mix of incomes and social groups.

In 1995, however, the newly elected Conservative government cancelled more than 300 planned housing projects (cooperatives and non-profits) while cutting day care, health care, legal aid, and support for tenant advocacy groups, and reducing welfare payments an average of 21.6 per cent.

The principal programs in the province of Manitoba are the Winnipeg Core Area Initiative and the Main Street Project. The former is a five-year $96 million program funded equally by the federal, provincial, and municipal governments. Intended to achieve comprehensive socio-economic development in the downtown area, it includes the rehabilitation of 4,000 dwellings utilizing existing public and private non-profit housing agencies.

The Main Street Project is located in Skid Row, which has an unemployment rate over 50 per cent and an incidence of family poverty five times the average for the city of Winnipeg. This district has the greatest concentration of poor housing and the highest number of Natives and ethnic immigrants. Main Street began as an emergency overnight shelter and daytime drop-in center for the unemployed. Subsequently it expanded to include a crisis intervention and emergency referral service, street patrols to assist homeless people, a detoxification unit, a case management and assessment program, and a day treatment center. The program operates a twenty-six-bed hostel for the temporarily homeless and for people entering or leaving treatment programs and is planning an eighty-eight-bed "supportive housing facility." Main Street deals with about 7,500 individuals annually. About 42 per cent have no fixed address. Two-thirds are Natives or Métis.

In Nova Scotia, the Province concentrates on programs which help welfare recipients to become home-owners. Social assistance payments can now be considered as income in order to meet qualification criteria for eventual purchase of dwellings. Mortgage funds are committed which, along with a second mortgage program, provide for 100 per cent financing of new or existing units.

In Quebec a $15.5 million federal/provincial program has developed permanent housing for homeless people. It is expected to result in the production of 260 homes with 1,300 dwelling units by non-profit housing organizations. A $2 million provincially funded program supplies equipment (principally in kitchens) for existing housing to assist disadvantaged groups. In addition, arrangements are made so that homeless people without a fixed address can receive welfare payments (Daly 1990).

THE RESPONSE TO HOMELESSNESS BY MUNICIPALITIES IN CANADA

During the 1970s and 1980s it became clear that the private sector in Canada was unable or unwilling to provide rental housing for low-income groups. At the same time, it was generally acknowledged that no more large, monolithic public housing projects, accommodating only very low-income households, would be built. Most municipalities and some provinces attempted to fill this void by developing, facilitating, and/or funding social housing.[7] A lead role was taken by large municipalities; they have the power to zone, regulate land use, and to establish building, development, and subdivision standards.

There are several models for municipal involvement in the production of social housing: one is the municipal non-profit housing corporation, like Cityhome in Toronto, which developed 400–500 units per year during the 1980s. Cityhome's guidelines define those in need as low-income households in crowded, inadequate or unsuitable dwellings, or those low-income people who are paying more than 30 per cent of their income in rent.

A second model is employed when the city assumes a facilitator role to support the development of social housing. Vancouver uses this approach to assist developers, by providing land and subsidies, by streamlining permit processing, and by lobbying senior governments for funding (Carter and McAfee 1990: 233).

Another approach, where there is an active voluntary sector, is for the municipality to provide funding for non-profit housing corporations and co-operatives which develop social housing. These initiatives, now the principal means of providing affordable housing in Canada, rely heavily on provincial and federal funding. Other municipal efforts to increase or maintain the stock of low-cost rental housing include residential intensification in cities, rehabilitation, down-zoning, and inclusionary zoning or zoning bonuses.[8] Residential intensification reduces commuter traffic and makes better use of the existing stock by allowing basement apartments and permitting conversions to include apartments or granny flats, additions, and home-sharing arrangements. This approach usually requires the formal or informal relaxation of municipal zoning and building requirements. In some Canadian cities with tight housing markets, as many as one-third of the single family homes include rental units.

Some rehabilitation efforts are directed at improvement of SROs to ensure that existing rooming houses do not deteriorate beyond the point of usefulness. Other programs are designed to assist low-income home-owners. In addition, non-profit housing corporations rehabilitate older dwellings for occupancy as group homes serving low-income singles. This approach makes use of buildings in poorer districts where it is unlikely that development will provoke neighborhood protests. Non-profit groups, however, run the risk of failing to meet fire, safety, and building codes. Officials usually enforce these regulations more rigidly in group homes than in single family residences.

Down-zoning has been employed to preserve inner city neighborhoods. Public intervention has been successful in conserving older buildings and preserving the physical fabric of certain districts, but unsuccessful in retaining the socio-economic character of neighborhoods. In most cases, down-zoned districts ultimately become gentrified by affluent newcomers attracted by the well-preserved residential buildings (Ley 1993).

Inclusionary zoning and bonus systems are used in a number of North American cities. Municipalities require developers of commercial projects to provide affordable housing as a trade-off for development approval or for bonuses in the form of increased floor area allowances. These programs work best in cities, like Vancouver, Toronto, Boston, Seattle, and San Francisco, where there is limited developable land and (in recent years) pent-up demand for office space.

COMPARISON OF GOVERNMENTS' RESPONSES TO HOMELESSNESS

The following is a comparison of the British, American, and Canadian approaches to homelessness in terms of the key questions suggested at the beginning of this chapter.

Relationships between central and local governments

Britain's political economy is characterized by class divisions and regional disparities; increasingly, political matters also depend on ethnicity and ideological tensions. The role of each level of government in recent years has been dictated by economic restraint, by the Conservative government's objective of reining in local authorities, and in substantial measure by the ideological fervor associated with privatization.

Inter-governmental politics in the United States are pluralistic, often dominated by race or parochialism, or both. Even though cities, as creatures of the states, lack significant legal powers, they have considerable latitude in determining how, where, and to whom federal dollars are dispensed. Resistance to federal interference is common and there is a presumption of local autonomy in what many regard as a local issue. Until 1993 federal authorities failed to provide either carrots or sticks to local government in order to preclude homelessness.

As a federal state there are tensions among the three levels of government in Canada with respect to the provision of shelter. Debates are commonplace over who will provide the funds and who will receive credit for creating new dwellings. Although the federal government has devolved authority to the provinces there is a great deal of variation in social housing production in different parts of the country. The provinces allow municipalities substantial room to develop social housing. Results are mixed, depending on perceived needs and the political leverage exercised by housing advocates. With the introduction of the deficit-cutting budget in 1995, the Liberal government declared that the federal government would become a minority – and silent – partner in maintaining the social safety net; the responsibility for meeting society's health, education, and welfare needs now rests squarely with the provinces and municipalities, which are being squeezed by the pressures of fiscal restraint, lack of growth in tax revenues, and reductions in transfer payments from the federal government.

Effectiveness of public policy

British government policy appears to be effective when there is a genuine local commitment to house homeless households, to assist those in priority need who are at risk, and when sufficient resources can be made available for adequate shelter. But the links between housing and other social services, training, and employment programs are not yet effectively coordinated.

In the United States prior to 1987 no comprehensive public policy with respect to homelessness existed at the national level. The McKinney Act incorporated a number of useful programmatic concepts and acknowledged the need for federal intervention. Though underfunded in some key areas, this legislation represents an important step in recognizing homelessness as a national problem. With the ascendency of neo-conservative Republicans in the mid-1990s, however, funding for these social programs is jeopardized.

Federal policy in Canada encourages the development of social housing by other levels of government and by the voluntary sector. This policy is effective when provincial and municipal (or regional) governments are motivated to mount an aggressive housing program. Without this local commitment the policy goals cannot be realized. Moreover, Canada has no comprehensive national policy to link housing and supportive services with programs intended to address homelessness.

Gaps between policy and practice

This is a crucial issue. The variation among authorities indicates that national policy in Britain is substantially altered and, in some cases, ignored by unsympathetic local officials. The capacity of local councils to respond is hampered by lack of resources.

Execution of policy has been fragmented in the American federal system. It is difficult to develop comprehensive long-term responses when homelessness is seen by many public officials as a short-term local phenomenon and funding is directed at emergency shelter rather than toward permanent housing. Some progress is being made, however, in response to local initiatives and implementation of the McKinney Act.

While Canadian housing policy is designed to create social housing for those in deep core need, production in recent years has been only one-half of the level of the early 1970s. Policy implementation requires substantial public funding and depends on the political will of provinces and municipalities. This is proving to be problematic as the federal government (in the mid-1990s) is concentrating on deficit reduction and budget cutbacks.

Advantages of the various approaches

Because British law includes those at risk, people who are homeless are placed in temporary or permanent housing and a great deal of attention (from local authorities and voluntary organizations) is focused on vulnerable groups, including low-income single mothers. A number of agencies have established tentative links among education, training, health and employment programs, and permanent housing.

Most of the creative projects adopted to address homelessness in the United States have been devised at the grassroots level. Usually these involve non-government groups or public–private joint ventures. A number of self-help initiatives and joint projects offer considerable promise.

Perhaps the single most important characteristic of the Canadian approach is the existence of a relatively comprehensive social safety net and the provision of free health care for poor people. Another element of the Canadian system (at least in Ontario and Quebec, which represent more than 60 per cent of the total population) is that single people may receive welfare assistance even if they lack a fixed address. In some cities public and private agencies are moving away from the traditional model of large shelters and giving more attention to small group homes, transitional housing, and permanent dwellings on scattered sites.

CMHC's decision to devolve authority to the provinces resulted in an increase in the production of social housing by municipal non-profit housing corporations and private non-profit groups. These mixed-income projects function reasonably well and represent a major improvement over monolithic public housing projects of earlier decades. By sponsoring or funding non-profit or social housing CMHC and provincial governments have fostered the development of some projects which are planned, designed, and managed by users. Post-occupancy evaluations reveal that these developments achieved most of their objectives and are functioning reasonably well.

Disadvantages of the various approaches

The British place undue reliance on voluntarism, privatization, and devolution to the local level. Government policy directives are not always supported by transfer of adequate resources to local authorities, whose power was circumscribed at the same time as they assumed greater responsibility for homelessness. Without a committed local authority it is difficult for homeless people to secure their rights. Immigrants have considerable problems with the highly bureaucratized housing and social service systems. Frequently, the housing provided by local authorities in Britain is poor. Critics note that guaranteed housing provision represents "an extension of rights of citizenship . . . [but] it is of little value unless accompanied by wider housing policies which result in actual provision of accommodation which is adequate in terms of quantity and physical standards, and is appropriate to the needs of those to be housed" (Anderson 1993: 18). Housing alone may be insufficient to address the problems of very poor households especially among refugees and ethnic minorities as well as lone parents. Employment opportunities, which determine income and affect housing options, are extremely limited.

The American experience, in which ameliorative measures are often temporary, fragmented, or misdirected, demonstrates both the effectiveness of charitable or voluntary endeavors – and their inherent limitations. The record to date in the United States also provides evidence that privatization often does not benefit the poorest, who are left behind in the rush to downsize government and foster self-reliance.

In Canada, while federal subsidies are limited to the core needy, provincial and municipal housing subsidies are directed at moderate-income households as well as those in deep core need. This has the effect of "de-ghettoizing" developments but also reduces the amount of assistance available for the poorest households. There is constant tension between the need to erect new units and the necessity of winning community acceptance for subsidized housing. Public authorities and third sector providers often opt to build social housing in poor urban areas where residents have little political power. During the 1970s and 1980s gentrification and residential intensification caused inner city land values to escalate. Suitable sites for social housing are now difficult to acquire at affordable prices. In addition, delays inherent in the processing of planning and development applications (because of stringent codes and community opposition) inhibit the production of social housing and make it more difficult to deliver low-cost units. Non-profit and cooperative producers assumed some of the responsibility previously taken by government for delivery of affordable housing. This arrangement worked well in many instances. The housing produced is well designed and built to a high standard, but it is expensive, and, inevitably, controversial (Donovan 1995: A1). In order for production to continue government funding must be assured. Recently, cuts have been effected and the level, duration and security of such support is precarious.

The legislative responses to homelessness in the three countries are summarized in Table 12.

SUMMARY

All three countries have experienced a wave of neo-conservatism during the 1980s and 1990s at the same time as homelessness increased and affordable rental housing became scarce. Britain has been under Conservative rule since Prime Minister Thatcher's election in 1979, followed by John Major in 1990; the United States elected Reagan in 1980 for two terms, followed by Bush (until 1992); Canada elected the Progressive Conservatives (Mulroney) in 1984, and they remained in power until 1993. The three countries experienced deep, prolonged recessions in the early 1980s and early 1990s, which led to cutbacks in social spending and tightening of eligibility criteria for welfare benefits. Experiments with workfare and welfare restrictions continue, along with efforts to further privatize health and social services.

Despite these similarities striking differences are evident in the approaches taken in Britain, Canada, and the United States. British local authorities are required to accommodate homeless people in priority need, though regional differences in implementation are pronounced. The federal government in the United States enacted legislation to address homelessness in 1987 but there is still a laissez-faire approach with limited public involvement except in the provision of emergency shelters. No national legislation exists in Canada to contain homelessness but efforts to provide permanent social housing have been relatively successful in accommodating a range of income groups, including the poorest.

Table 12 Comparison of legislative responses to homelessness in Britain, the United States, and Canada

	Britain	*United States*	*Canada*
Legislation	The Housing (Homeless Persons) Act of 1977, as amended, supplemented by Code of Guidance	Stewart B. McKinney Homeless Assistance Act of 1987, as amended	No federal legislation on homelessness; rather, there are federal, provincial, and municipal acts and programs to provide social housing and financial assistance to those in "core housing need"
Responsible agencies	Department of Environment and local authorities as well as housing associations	A number of federal agencies administer programs; grants made to states and voluntary agencies	Federal, provincial, and municipal governments; grants transferred to non-profits and cooperatives
Definition of homeless	Has no secure accommodation; separated from rest of household; subject to violence or threats. Priority given to dependent children, pregnant women, and those vulnerable due to old age or disability	Initially defined as those on streets and in emergency shelters; later expanded to include welfare hotels and some of those at risk	Those without adequate, secure housing and those in core housing need who require shelter plus other forms of assistance
Nature of accommodation	Local authority obliged to house if in priority need *and* homeless; use council housing, hostels, and bed and breakfasts. Supposed to provide permanent housing	Emergency shelters opened by municipalities and voluntary groups; big cities rely on welfare hotels and motels	Some reliance on emergency shelters; provision of permanent and transitional social housing by public and private non-profits as well as housing cooperatives

Several themes emerged with respect to the effectiveness of public programs. First, the way in which homelessness is defined by officials frames the public sector response. In general, central governments see the homeless population as a local problem, involving a virtually powerless constituency, which can safely be ignored or delegated to the voluntary sector. Many of the actions taken at the national level in Britain, the United States, and Canada, served to exacerbate homelessness: by abdicating responsibility for publicly assisted and low-rent housing, by cutting spending on social services, employment and training programs, and by reducing the numbers eligible for social security and welfare benefits.

Public sector responses emphasize emergency shelter and temporary programs, reflecting a belief (or hope) that the problems associated with homelessness will be short-lived. Government measures lack continuity; this is not unexpected, given the whimsical nature of the political process and the need for an issue to generate considerable public interest in order to gain the attention of officials.

There is often a substantial gap between policy and practice because of a lack of resources, an inadequate supply of affordable housing in locations where it is needed, variations among local authorities in interpreting their legal duties, and a tendency for their responses to be shaped by ideology and parochial concerns.

The most important functions of national governments are to formulate policy and to provide funding for programs which will be shaped to community needs and carried out at the local level. In many cities, where public officials are sympathetic, a comprehensive range of innovative projects exists as a result of cooperative efforts between public and voluntary agencies.

Because of differences in ideologies and in measures taken by government, the nature of the key actors changes from place to place. In particular, the role of the voluntary sector is important and there is a trend toward public–private joint ventures to produce housing. These changes and the roles of voluntary agencies are explored in Chapters 10 and 11.

10

CHANGING VIEWS OF CHARITY

The third sector's role in addressing homelessness in Britain, the United States, and Canada

[A] society in which the vast majority of men and women are encouraged and helped to accept responsibility for themselves and their families . . . who care for others and look first to themselves to care for themselves.

(Prime Minister Margaret Thatcher 1977)

To a lot of people who give to charity, or who have never needed it, charity is a good thing. It's sharing . . . but when you don't have anything to eat for yourself, let alone donate, then you see things differently . . . it's humiliating and degrading.

(World Food Day Association 1993: 4)

Who should address the issue of homelessness? How have voluntary agencies in Britain, Canada, and the United States responded to the problems associated with homelessness? Voluntary groups assume a number of roles: some are effective lobbyists and shapers of public policy; a few fulfil an essential education function; many are accomplished advocates and mobilizers of grassroots support; others are adept at involving users in planning and managing their own projects. In recent years the third sector's role has been enlarged as the social welfare function has become increasingly privatized.

Public sector activity is concentrated on grants, mortgage guarantees, and other forms of financial assistance to the voluntary or third sector. Most non-profit groups cannot do an adequate job without government funding; generally, however, they are better at operating local programs than are public agencies. Many, like church groups, own their facilities outright and are not burdened with capital expenditures or mortgage carrying costs. Most rely at least in part on contributions (food, money, and goods), on unpaid labor, and operate more economically than government bureaucracies. Small, flexible, community-based organizations may be better equipped to understand the needs of homeless people and devise appropriate programs while adapting to changing circumstances and to the needs of different populations. Voluntarism, though, is not a panacea.

39 Nurse's Settlement House (ca. 1900) provided a playground for immigrant children in New York City (U.S. National Archives)

THE NATURE AND FUNCTIONS OF THE VOLUNTARY SECTOR

The voluntary sector – also known as the independent, or non-profit sector – is distinguished from the public (i.e. government) sector and the private (i.e. business) sector. Organizations within this third sector are classified as those relying entirely on government funding (quasi-governmental bodies), those which depend on both public funding and fees or contributions (private service agencies), and those which are entirely voluntary (or cooperative) and do not receive government aid. In Britain these agencies are known as non-statutory organizations while in North America they may be called NGOs (non-governmental organizations) or CBOs (community-based organizations); the former are typically larger and are registered or incorporated; CBOs are usually small, informal groups which evolve out of a grassroots effort to address a specific need. As used in this context, voluntary agencies are primarily concerned with social services. Some provide care; others are advocates, educators, lobbyists, publicists, or catalysts for change. Many large agencies undertake several of these roles (Tobin 1985).

Among the functions ascribed to voluntary organizations are: initiating new ideas, especially in areas where public agencies lack experience; developing public policy through research, by litigation, or by means of pressure exerted on politicians and public officials by representatives of social, political, environmental, or economic movements; supporting minority and local interests; providing a forum for citizen participation; acting as a catalyst to stimulate or reinforce activities of government and business where they interact with voluntary agencies; monitoring government and the marketplace to assure that public interests are served; initiating legal action, if moral suasion fails, to ensure that public and private agencies comply with the law (Commission on Private Philanthropy and Public Needs 1975: 41-46).

The third sector includes a broad array of institutions, ranging from small informal grassroots organizations to international multi-purpose agencies. In this discussion they will be treated in a rough hierarchy ranging from missions, soup kitchens, food banks, and emergency shelters at one end of the scale, to permanent housing programs linked with comprehensive social services. In between are many levels of voluntary enterprise. For the sake of simplicity these are broadly defined as advocacy, housing, counseling, social service, and training schemes. A range of alternative programs and approaches is found in each of these areas:

Advocacy Groups advocating changes in public policy and practice use such tactics as fasting or litigating in order to focus attention on homeless people and their living conditions; projects range from health promotion, to legal aid, to citizens advice centers.

Housing In addition to emergency shelters, projects include hostels for groups like teenagers or battered women, single room occupancy (SRO) lodgings, half-way, transitional or group homes (through self-help, cooperative or non-profit initiatives), and various schemes to produce permanent dwellings.

Health and social services Approaches include day care, transportation, referral and linkage with community service agencies, mobile or street clinics, as well as attempts to assure equal access to housing, health care, education, and employment.

Counseling, advice, and information Projects include drug and alcohol detoxification and counseling, information and referral centers, and rehabilitation programs.

Education and training Programs range from basic "life skills" (e.g. budgeting, child care, literacy, housekeeping) to job training schemes or diploma programs leading to apprenticeship or permanent employment.

What is the actual experience of these groups in dealing with homelessness? The record is mixed. Many charitable organizations have experienced a metamorphosis. They began with a mission: to save the homeless person through spiritual uplift, education, and demonstrations of kindness. Workers in these agencies initially believed that they were dealing with a short-term problem which could be solved with zeal, voluntarism, and faith, along with such measures as emergency shelters, soup kitchens, and food banks. Their experience is reminiscent of the settlement workers and friendly visitors of the late nineteenth century. While some "rescue missions" still pursue an evangelical approach, many care-givers now realize that, despite their best efforts, the level of homelessness has not declined. They struggle with the dilemma of cajoling governments to initiate long-term programs while their voluntary groups try to cope with daily emergencies and lean budgets.

Advocacy: who speaks for whom?

Virtually all third sector organizations engage in some form of publicity, education, fund-raising, or advocacy. There are, though, a few groups whose primary mission is to serve as advocates on behalf of homeless people. Their work includes research, litigation, lobbying, political action, and public education to inform people about homelessness and housing conditions.

Associations like the Coalition for the Homeless in the United States, the Canadian Council on Social Development, and several British organizations, notably Shelter and CHAR, are adept at statistical and qualitative research designed to focus public attention on living conditions. They conduct surveys or censuses, publish periodicals, produce documentaries for television, and provide visual and statistical information for the press. Related groups, like the Children's Defense Fund and the Child Poverty Action Group, perform similar functions for particular segments of the population who lack representation.

Some of these organizations have multiple functions with respect to the public sector: they act as watchdogs to hold agencies accountable, to ensure that public bodies perform according to mandate and that peoples' entitlements are honored; on a regular basis, they provide testimony to legislative committees and public hearings. Others, like the Low Income Housing Coalition in the United States, concentrate their energies on shaping public policy and on bringing pressure to bear on legislative committees.

Another type of association is active in filing class action suits on behalf of homeless people. Litigation is designed to secure legal recognition for homeless people and to ensure that their right to shelter is guaranteed. In New York and Washington, D.C., litigation has succeeded in ensuring that the entitlements of homeless people are honored by public agencies and shelter providers (Kanter 1989: 91–104). Groups like the Community for Creative Non-Violence have been active in this area and have galvanized public attention by publicizing fasts, marches, squats or "occupations," and other events intended to attract media coverage. Notable examples in American cities during the late 1980s and 1990s are the squats and occupations at Thompkins Square Park in New York, the Love Camp in Los Angeles (Rowe 1988: 16), and Tranquility City in Chicago.

Changes in strategy are evident. Some organizations effected a transition from public demonstrations to provision of shelters once they succeeded in alerting citizens to existing problems. Others are moving beyond hostels to develop permanent housing; and certain groups, like the Daily Bread Food Bank in Toronto, are taking on both political and business attributes. Representatives of the food bank place considerable importance on regular press stories designed to educate citizens and politicians on the need for involvement of public agencies in providing for the food and shelter needs of their homeless clients. Recently, they have also examined their organization from a business perspective – perhaps implicitly acknowledging that their goal, to work themselves out of a job, has failed. Graduates of the Harvard Business School are advising the food bank on marketing strategy, organizational effectiveness, and ways to improve efficiency.

40 Demonstration by the Lower Broughton Housing Action Group against their council's failure to move them into decent housing (Photograph courtesy of Shelter)

Squatting

Perhaps the most common form of grassroots organizing is squatting. A San Francisco group of homeless people, Homes Not Jails, has coordinated squatting in abandoned buildings in a covert and, in some cases, an overt manner designed to gain exposure and enhance public awareness (to illustrate, for example, that there are more than 6,000 vacant buildings or houses in San Francisco). One of the group's founders, Ted Gullickson, explains this process:

> After you help someone get a place to live, they see you are really doing something. In most homeless services there is a very clear distinction between well-paid staff people and their clients, who they see as needing their paternalistic help to get off the street. What we are saying is that we all have a right to a home, and people respect that . . . Homeless people understand this message. They can understand the connection between private real estate and poverty. They are angry at being marginalized. We help them channel that anger into action.
>
> (Mesler 1995: 2)

These examples indicate that the response to homelessness has become institutionalized in some settings. Moreover, most advocacy efforts on behalf of homeless people are initiated by middle-class, well-educated professionals who may not have anything in common with their "clients." Some critics allege that their efforts, particularly those that deal with symptoms by providing short-term cosmetic solutions, in effect cause the advocates to be part of "the homelessness industry" which may perpetuate the problems that it ostensibly is working to eradicate. Paul Boden, the formerly homeless director of the Coalition on Homelessness in San Francisco, believes that "the problem is that so many groups are totally disconnected from the base they are supposedly representing. These groups are poverty overseers. What is killing poor people is that they are not being allowed to create their own agenda" (Mesler 1995: 5).

As a result of such criticism, more attention has been directed, of late, to authentic grassroots efforts and political activism which are directed by homeless people. These range from newpapers sold by homeless people on the street corner, to public education campaigns, to e-mail networks on current events relating to homelessness (e.g. legislation), to computerized searches for missing street people and runaways. Other efforts involve litigation, sit-ins and organized squats. The Coalition for the Homeless in San Francisco, for example, includes black lone parents and a Latino advocacy group (*Ayuda*) working to make people aware of the extent of homelessness among such individuals (many of whom are virtually invisible); they also intend to challenge the institutions with a role in perpetuating homelessness. Another is Food Not Bombs, with chapters in sixty-six cities across Canada and the United States. It serves food to homeless people in public places, often in front of city halls, to confront politicians and to challenge laws which prohibit food service in public without having a permit.

41 Occupants of B & Bs in Bayswater, London, demonstrate against homelessness and the conditions in temporary accommodations.
(Photograph by Jonathan Stearn, courtesy of Shelter)

An argument for self-advocacy

I am a former client who still lives in a SRO supportive environment. I really appreciated the title change [from "case manager" to "service coordinator"] when it occurred at my agency. It gave me a sense that I was no longer being managed, and resulted in my taking a more active role in my treatment plan and subsequent recovery from homelessness, substance abuse, and mental illness. The increased sense of having some control in my life led me to conduct self-advocacy for the removal of client status. I am now 16 credits short of my B.A. and have worked for the last three years as a Librarian and an Internet Specialist for the agency where I was a client . . . The future seems a heck of a lot brighter today than it did 6 years ago.

(Kevin Childs, August 4, 1995)

The Coalition also includes StreetWatch, which videotapes police actions against homeless people. The police issued thousands of citations (particularly to homeless people sleeping in a public park) as part of the Matrix program, an initiative justified by Police Chief Anthony Ribera in the name of crime prevention: "We need to stop the little crimes, the peeing in public, the public drinking, and the big crimes will stop."

Food banks and soup kitchens

A social indicator of the intractability of homelessness is the proliferation of food banks, soup kitchens, and clothing depots. Soup kitchens were ubiquitous in the 1930s and many people would not have survived the Depression were it not for the daily sustenance they provided. They were operated by groups as diverse as missions, church ladies' guilds, and Al Capone's organization in Chicago.

Typically, food and shelter are provided by a mix of public and voluntary agencies; the latter usually receive government funding to defray expenses. In the United States during 1987 a total of 321,000 meals were served each day to homeless people in 3,000 shelters and soup kitchens, most of which began operations during the 1980s. Almost half are shelters that serve at least one meal daily, and 41 per cent are soup kitchens, three-quarters of which serve one daily meal (Burt 1992).

Reliance on these community institutions operated by missions, churches, and volunteer groups has increased sharply. No food banks existed in Canada in 1980, but by 1993 the Canadian Association of Food Banks had over 300 members, supplying over 1,200 grocery programs and 580 outlets that dispense prepared meals (Canadian Association of Food Banks 1993). Food is donated by individuals and by farmers. The major source, however, is the food industry – including brokers, processors, retailers, manufacturers, and hotels. Goods that have been mislabeled, are approaching expiry dates or about to be discarded are used by food banks. This practice will become widespread if "good samaritan" legislation is passed which will relieve providers of liability

42 Soup kitchens were operated by a variety of groups during the Depression. Al Capone's organization in Chicago offered free soup, coffee, and doughnuts to the unemployed, 1931 (U.S. National Archives)

if recipients become ill from eating tainted or spoiled food. In the absence of such laws, producers and retailers are reluctant to contribute essential but perishable items (like meat, fish, and dairy products). As a result, the offerings of most food banks are limited to canned goods and other non-perishable items. A Government of Ontario report in 1990 found that "in most cases, food banks are unable to ensure an appropriate balanced diet" (Government of Ontario 1990). Because of high demand about 90 per cent of food banks in Canada must ration their goods: frequently, users are limited to one bag per week.

In 1991 more than 2 million people (roughly one in twelve Canadians), including more than 700,000 children under age 18, received food assistance at least once. About 590,000 people each month accepted help from food banks, representing an increase of more than 50 per cent from two years earlier (Statistics Canada 1992: 12). Two-thirds of those who customarily obtain their food from emergency meal programs and food banks are welfare recipients, 12 per cent cite employment as their main income source, 6 per cent rely on old age pensions, 4 per cent on unemployment insurance, and 3 per cent on disability pensions; 4 per cent have no income (Statistics Canada 1992: 6–14). A November 1992 survey by Montreal Harvest found that about one in four people using food banks and soup kitchens is doing so for the first time.

In Metropolitan Toronto food banks were patronized by 35,000 people monthly in 1986, 93,250 per month in 1990, and by 120,000 per month in 1991. The Daily Bread Food Bank in Toronto, Canada's largest, distributes food through a network of 235 programs. Demand in 1991 increased by 44 per cent over 1990, while donations grew by only 20 per cent. Relying on contributions and a municipal grant, the Daily Bread remains open seven days a week, has ten paid employees and about 500 volunteers working in an 18,000 square foot warehouse and a 40,000 square foot storage facility. Although it relies on frequent food drives, 70 per cent of its donations come from the food industry. Similar patterns are evident in other Canadian cities. Montreal Harvest serves 170 agencies which stock food for 53,000 people each month; in 1993 it distributed 20 tons of food daily. Roughly half of the recipients are under age 18. Food bank operators have serious concerns about the trend toward institutionalization of their services. Repeated attempts have been made to induce governments to introduce welfare changes which would render the banks obsolete. The only response to date has been in Ontario, where the provincial government in late 1990 established a $1 million fund to relieve pressure on food banks and injected $300 million in the social welfare system to reduce food bank usage. Because of increased demand, however, food bank usage continued to grow.

The largest supplier of food to poor people in the United States is Second Harvest, which accepts donations from corporations, distributes these to food banks, and charges a fee (0.0088 cents a pound in 1995) to cover shipping, handling and administrative costs. Unfortunately, Second Harvest receives a great deal of inappropriate or "dumped" goods which it cannot use, but which net the corporate donors tax write-offs (*Chronicle of Higher Education*, March 23, 1995: 42).

Community kitchens and food banks serve immediate needs. But they also present problems. Often relying on sporadic in-kind contributions and a tenuous pool of voluntary labor, they exist from day to day and deal with symptoms of poverty, rather than with underlying causes. Some believe that they perform functions which should be undertaken by government. Others concede that, in the absence of public sector intervention, they are essential for those who lack alternatives.

Shelters

Much the same can be said for emergency shelters. Some of these facilities are little more than warehouses for people: they lack beds and only provide sleeping space on bare floors. Others have cages of wire mesh and offer no privacy or relief from pervasive noise and smells. Many operate close to capacity year round and regularly turn people away because they are over-booked. A survey in London found that hostels were at 87 per cent occupancy. Studies in Canada have yielded similar data (Canadian Council on Social Development 1987). Some advocates argue that shelters are essential for the immediate future, while others believe that they perpetuate homelessness by deflecting attention from the need for permanent housing.

A profile of people using food banks

"What alternatives do you have? If you don't have them, you can't use them, and you use them because you need them."

"If there was some other resource, we'd just say forget it. We wouldn't continue to come here and be put down. I think that people end up feeling very, very demeaned, and there's women that don't come anymore and I wonder . . . how they survive."

"No one else out there is going to do it for us. If the food bank closes tomorrow, as someone said, I'd beg, borrow, steal, sell my body to feed my children."

"What everyone is trying to do here is better their families. And the fact that we don't have any choices is [because] number one, we're women. We make substantially less in the work force than men do. The fact that we're single women, that's another strike against us, and that we have children."

"Some of the things I get here have expired dates or they're not really good for you. I just want to be treated like another human being and have ordinary food that tastes good and is good for the kids too. All of these chocolates and candies. The kids love it, but it's not healthy. I open up the bread and I find this blue stuff on it and I don't use it."

"I have four children. You have to wait in line in a basement for an hour with four children for a bag of food. It's frustrating. The one I was going to is dark and dirty and full of adults."

Critics of food banks believe that they have become a secondary welfare system which provides "second-rate food to people who have been obliged to become second-class citizens . . . and the very existence of food banks sustains this."

Karen Shaver, formerly associated with Toronto's largest food banks, feels that they meet many needs: "the need of people to help, the need of industry to get rid of its surplus and waste food, and the needs of people who use them. And the last one is not necessarily the most important."

(World Food Day Association 1992)

In the United States, about two-thirds of emergency shelter operating expenses are covered by government grants; state and local governments contribute between half and three-quarters of this total. Private sources provide the remaining third of the shelter operating funds and the great majority of the staffing services. It is common to find a mix of funding and staffing under one roof. The Long Island Shelter in Boston, for example, receives state and city funding as well as federal monies under the McKinney Act. Services are directed by paid staff, but most services are provided by volunteers, particularly during meals when over 300 people must be served within a short time.

Certain organizations have a vested interest in maintaining the poor. Among the shelter providers are traditional missions whose role is to salvage lost souls. In many cities and towns, the mission is the only shelter operator. Some, though not all, require that guests attend sermons and religious services in exchange for lodging. These institutions typically have rigid rules, will accept only men, and offer only basic accommodation. Shelters run by non-religious groups tend to be more flexible than traditional missions and are operated by both volunteers and professionally trained staff. The majority

43 Women from a shelter participate in a communal kitchen, 1995 (Photograph courtesy of Joyce Brown)

receive some public funding, but are generally smaller, less bureaucratic and more cost-effective than their municipal counterparts. Virtually all, however, suffer from a chronic shortage of resources. Because they are dependent on contributors and government grant agencies, some shut down periodically when their budgets are depleted; others have closed permanently for lack of funds.

Missions

Francis slept in the weeds on Dougan Avenue below the bridge and woke up about seven o'clock this morning, then went up to the mission on Maddison Avenue to get coffee . . . Francis had coffee and bread with the bums who'd dried out, and other bums passin' through, and the preacher there watchin' everybody and playin' grabass with their souls. "Never mind my soul," was Francis's line, "just pass the coffee . . . He puts out a good meal, though . . . He does in a pig's ass . . . Wonderful . . . Pig's ass. And he won't feed you till you listen to him preach. I watch the old bums sittin' there and I wonder about them. What are you all doin', sittin' through this bullshit?"

(Kennedy 1983: 6, 7, 22)

Transitional housing and group homes

Women's hostels, refuges and half-way houses often were founded by groups with a commitment to aid battered women. Organizations like the Women's Institute for Housing and Economic Development in Boston, Massachusetts, moved to second stage or transitional housing to accommodate women who are ready to leave shelters but are unable to find suitable lodgings (Sprague 1991). Transitional projects are small, often with communal areas, and residents are permitted to stay from six months to two years. They provide women with a bridge to self-sufficiency and permanent housing. Some offer counseling, life skills training, child care services, and job preparation or referrals. Others assist residents in paying rental or security deposits when they move on.

The typical group home format consists of five or more individuals, such as former substance abusers or runaway youths, sharing a house which has private bedrooms but communal living and dining areas. Some serve as half-way houses for people when they leave prison, detoxification centers, or psychiatric institutions. Group homes are used most frequently for recovering alcoholics and drug abusers; residents subscribe to house rules which require that they be "clean and sober." Violations lead to expulsion. While there are problems with this model (and many take issue with its obligatory nature) it often works for many people because of peer pressure, a supportive family-like atmosphere, and the fact that all residents have similar experiences.

There are difficulties in establishing group homes, however, resulting from community resistance. Where opponents are well organized and able to exercise political leverage, group homes have been excluded. Resistance is strongest in suburban areas; arguments that group homes should be equitably distributed throughout metropolitan regions have not been persuasive. Consequently, homes most often locate in inner-city neighborhoods which are

Group homes for recovering alcoholics

A former resident and manager of group homes concluded that:

> little else works for recovering alcoholics and addicts, at least a large portion. I am referring to the chronic and homeless who are taking the first steps back. For many, they have burned their social and financial bridges beyond real repair. Relatives have long given up on these people, due to the nature of the disease. Without a long-term family setting, these people will not have any safety net and safe environment to restore the needed basics . . . And the whole group continually learns one necessity of living, being part of the community, with its responsibilities, rather than lost on the street.
>
> (D. Nichols, May 18, 1995)

not an ideal setting for recovering substance abusers. Because of the "not in my back yard" phenomenon there is a shortage of facilities for young people. One organization, Youth Without Shelter, tried unsuccessfully for five years to establish a group home in metropolitan Toronto. After finally locating to the west of downtown Toronto, it opened a twenty-two-bed facility: in its first full year of operation, it was forced to turn away 2,472 young people because of lack of space.

In many cities group homes are prohibited by zoning codes. Ironically, in order to obtain federal funding, housing providers must comply with zoning regulations. This may be changing. In May 1995 the Supreme Court of the United States turned back to the lower court its ruling that zoning could effectively bar group homes from a single-family neighborhood (U.S. Supreme Court Case 94–23, *City of Edmonds v. Oxfordhouse, Inc.*).

Day programs

Homeless people have ample time on their hands. They need a fixed address during the day as well as in the evenings. Day centers accommodate them by acting as substitutes for hostels or shelters which discharge residents each morning. Some drop-in centers offer free meals or inexpensive sandwiches and coffee and a warm place to rest during cold or inclement weather. Others supply laundry facilities, free clothing, or counseling and referrals to health and social services. These facilities allow individuals to socialize in a relaxed setting which helps to avoid the debilitating effects of social isolation. Some places have a telephone and mail drop; these are indispensable for people seeking work or needing an address in order to receive welfare benefits. Many day programs lack adequate, regular funding. By providing a setting where health and social services can be offered to homeless people on an outreach basis, however, the day center serves a valuable function and is a potential model for the provision of community services.

Permanent housing

Most third sector organizations rely on short-term or emergency housing, because their energies are occupied in dealing with people in crisis. Under these circumstances it is difficult to think in terms of permanent housing. Moreover, the housing function of voluntary agencies is typically limited to shelter for specific client groups: alcoholics, ex-offenders, battered women, people with mental and physical disabilities, and young street people. This situation is gradually being altered, however, as events dictate reassessment of the third sector's role.

Some providers concluded that, given their experience with homeless people, they should develop and manage permanent housing. This judgment was reached after it became apparent in recent years that residents of their shelters and missions had become quasi-permanent occupants. Agencies realized that, even if satisfactory short-term housing were provided, their work would be futile unless affordable, permanent housing could be created.

This led to the emergence of private non-profit housing corporations, often managed by third sector groups that formerly ran shelters. Because of their tax-exempt status they can secure grants, government funding, and tax write-offs, thus providing a vehicle for low-rent housing production. In some cities the non-profits are complemented by municipal housing corporations and by housing cooperatives, managed by labor unions and by religious, ethnic, or other affinity groups (Daly 1990: 133–152).

Other services provided by the voluntary sector

Voluntary and non-profit organizations are concerned with the necessities: food, clothing, and shelter. The vast majority are not equipped to offer formal social services, counseling, or training. Moreover, some agencies believe that if extensive services are offered on-site where people are housed, residents will choose not to use the services already provided by regular social agencies. Many shelter users find that "linked" housing and rehabilitation programs introduce an objectionable connection that suggests compulsion. Accordingly, most third sector agencies provide referrals, some counseling and assistance with applications; beyond that, however, they require residents to get involved in daily programs (such as job training, counseling, and education) in order to avoid apathy and discouragement. Increasingly, voluntary organizations are becoming involved in training, to prepare people for employment and independent living, as governments in all three countries (as well as in Western Europe) are turning to a variety of obligatory workfare schemes for welfare recipients. The Homeless Action Coalition in various American cities provides homeless people with free access to the Internet (for job listings and for communicating with other homeless individuals and with service providers). Computers are also available for them to create resumes and job applications and to learn about jobs, programs, and services.

It is not enough, though, to insist that people are registered for programs and counseling. Indeed, some critics question whether it is appropriate to

44 Permanent housing in a completely renovated structure is provided by the Paul Sullivan Trust in Brookline, Massachusetts (close to downtown Boston) for former residents of inner city shelters
(Photograph by Jack Peckham for the Paul Sullivan Housing Trust 1989)

compel participation. Inevitably there are problems with coordination and with actual delivery of services to a diverse, often mobile, population. Problems include lack of coordination between different administrative agencies and ineffectual or non-existent linkages among such services as housing, health, welfare, and job training.

Because financial responsibility for programs frequently is shared by two or three levels of government, there are political and economic obstacles which may block effective coordination. While challenging voluntary groups, this situation enables them to play a role in bringing the parties together, by researching a particular problem, by publicizing their results, and by lobbying politicians. In the mid-1990s, however, conservative politicians sought to limit the lobbying role of non-profit groups in the United States. On August 4, 1995, the U.S. House of Representatives passed a bill (H.R. 2127) which included a provision prohibiting political advocacy activities by non-profit organizations that receive federal funds.

Though unwieldy and informal, oversight by the non-profit sector is necessary because the public system stresses legal and financial accountability. While reflected in policy statements or enabling legislation, social responsibility may be neglected in practice. Voluntary agencies (when they are not co-opted by their reliance on public funding) may serve as watchdogs on public administrators to ensure that their responsibilities are fully discharged. At the same time, however, third sector advocates stress the compelling need for more responsive administrative mechanisms to assure coordination of social services.

A complementary function is served by advocacy groups whose projects are designed to advise people of their entitlements and how to obtain them, to secure legal recognition for homeless people, to force public authorities to provide shelter as an entitlement, and to address basic causes of homelessness, including the lack of affordable housing (Kanter 1989: 91–104).

Since the 1970s a movement toward privatization has been under way in the United States; Britain and Canada are now following suit to some extent. Some Canadian provinces, notably Alberta in 1994/95, have adopted an American minimalist model of health care provision, seeking to cut costs and balance the budget (Rekart 1994). Critics allege that the ensuing cuts in service have been at the expense of long-term health (Armstrong and Armstrong 1994). Since 1990 (the National Health Service and Community Care Act) the British government has pursued a mixed economy of care model (in health, education, housing, and community care) which encourages a quasi-market relationship between statutory authorities ("financial enablers") and voluntary organizations ("providers"). The voluntary sector is quite concerned, however, about the possible loss of autonomy, the need to engage in a paper chase to secure contracts, the likelihood that agencies will "go where the money is" rather than meeting acute needs, and the conceivable decline in care standards if they should become based on short-term price competition (Lewis 1993: 189–191).

SUMMARY

Significant disparities exist between the public and the third sectors in terms of the size, nature, and objectives of the organizations involved. Voluntary groups themselves have disparate goals – as advocates, care-givers, publicists, lobbyists, trainers, counsellors, and innovators – in attempting to address needs not being met by government. They occasionally fulfill the important role of providing a forum for participation of citizens who wish to influence policy and decision-making. Non-profit organizations act as a buffer between the state, with its need to act expediently, and the concerns of individuals and interest groups in society.

Voluntary agencies, however, are heavily dependent on government funding. This precarious relationship endangers their stability, particularly during recessions, when their services are most in demand. Their need for autonomy conflicts with government's requirement for accountability. The organizational integrity of non-profit groups, as a result, may be threatened (Wolch 1990). If their continued existence is dependent on public funding it is highly unlikely that they will challenge controversial decisions, such as urban renewal plans which involve the demolition of SROs.

Both sectors are inefficient in some ways. While locally devised programs may be most appropriate, they do not necessarily result in ideal solutions. Typically, non-profit agencies are small, relying on low-paid staff and non-professional volunteers, subsisting on tenuous funding. Many of these organizations allocate a substantial percentage of time to fund raising and applying for grants. Though these groups are, in effect, a substitute for government in some situations, they rarely receive capital funds. Their efforts, understandably, often are hit-or-miss, ineffectual, or temporary in nature.

Certain voluntary agencies outlive their usefulness. Some focus on issues that are no longer relevant. Others continue to exist primarily to satisfy the needs of their leaders. A number of groups vie for the same territory – as well as for the same grants – but refuse to cooperate or consolidate services. Duplication, confusion and inefficiency result. Because of this complex web of motives and agendas voluntary sector activities may aggravate or prolong the problems they are designed to alleviate. As funding becomes more difficult to obtain it is increasingly common to hear complaints about the selection criteria used by these agencies to decide who will receive help. Accusations are heard about "cherry picking," when care-givers decide that only certain people (presumably the most likely to "succeed") are allowed access to services because there are insufficient resources to provide help to all applicants. This raises questions about who should decide, what criteria are used, and how success is defined.

Despite these drawbacks, a broad selection of useful projects is evolving as a result of innovators in the public and voluntary sectors. Several of these are described in Chapter 11.

11

ARCHITECTURE OF OPPORTUNITY

Innovative programs designed to address homelessness

> Housing and related problems have been cruelly over-simplified; there is more involved than the substitution of a clean box for a dirty one.
>
> (Elizabeth Denby, cited in Daly 1989: 412)

While projects cannot be hastily transplanted from one country to another, it is helpful to examine the successes and failures of others. The program review which follows for each of the three countries is based on a number of interviews that were conducted with national and local government officials, representatives of voluntary agencies, NGOs, community-based organizations, and program participants to assess programs intended to assist homeless people.

Since a diverse array of projects was selected for review, it was not feasible to apply the same standards of evaluation to all of them. Because the problems associated with homelessness vary with different populations, preference is given to programs that deal directly with certain groups of homeless people (ex-prisoners or young street people, for instance) or a specific issue (e.g. AIDS). Grassroots and self-help efforts (often involving joint public–voluntary ventures) are given particular attention.

Both short- and long-range programs are included, but most are relatively short-term in their outlook for several reasons. Many voluntary agencies exist on temporary funding and are characterized by rapid turnover and a relatively limited timetable; all of these factors affect project scope and goals. Some deal with people in crisis and lack the time or resources to engage in a more thorough examination of long-term alternatives. Also, long-range programs such as permanent housing represent a standing commitment which governments are hesitant to make when the future availability of funds and the status of voluntary agencies are uncertain.

The degree of participant involvement is an important criterion for selection and review. In the past a great deal was done for homeless people by well-meaning bureaucrats and care-givers. Frequently this approach induced dependency or hostility. Consequently, this review is focused in part on self-help or self-management schemes which allow people to plan their own housing and related programs and to exercise control over their lives. It appears that a key factor in determining the success of projects for homeless people is the degree of power and control they exercise. Included are economic power (access to resources and to the institutions that provide financing or goods and services), political power (expressed through the ability to control the project and to make decisions; it may also be expressed through political clout . . . the ability to make public officials take notice), and personal power (when the individuals involved are allowed to direct their own programs).

Another measure used in examining projects is the degree of comprehensiveness. Most homeless people have several problems simultaneously and need supportive services before they can manage housing and a job. Preference is given to schemes which link services, such as housing and education or training, so that participants can move on to permanent housing and employment. An additional consideration is adaptability because people's needs change over time. Finally, it is valuable to focus attention on programs which preclude homelessness or present alternatives to shelters and other temporary measures.

Despite the intricacy of the problems encountered in the three countries, and in spite of indifference by some public officials, a great deal of innovation has occurred. Progress has been made, particularly by the voluntary sector. Several of these creative schemes are discussed in the next section.

EXAMPLES OF PROGRAMS IN BRITAIN, THE UNITED STATES, AND CANADA

Information and advocacy centers

The concepts of power, control, information, and access to social benefits are related. Homeless people must know how to gain entry to a maze of service agencies in order to secure their entitlements. In Britain, virtually every city has a citizen's advice bureau and housing aid center which provide details on social benefits and services. Often these agencies are located in the same community center where people seek jobs, obtain social security benefits, and deal with local authority representatives.

Perhaps the best-known advocacy group is Shelter, the National Campaign for Homeless People, which has been operating in London and other major cities since the 1960s. Shelter, which deals with more than 62,000 inquiries annually, has an established reputation for conducting research and public campaigns on a variety of housing subjects and is a successful lobbyist for homeless individuals.

In London, advice centers are operated by London Shelter (formerly SHAC), established in 1969, which provides information on housing and related issues and accepts almost 2,000 advocacy cases annually. After London Shelter's intervention over 30 per cent of those who were previously denied help are offered council housing. Almost half of the agency's elderly clients are ultimately accepted by the boroughs following appeals.

Another voluntary organization in Britain, MIND, the National Institute of Mental Health's Housing Unit, advocates for people with psychiatric illnesses. MIND offers training programs to agencies which provide accommodation for people leaving institutions. Advice is given on finance, resident selection,

45 Counseling services provided at an inner-city shelter, Pine Street Inn, Boston (Photograph by Jack Booth 1989)

housing layout, support services, and residents' rights. MIND conducts campaigns to increase public awareness that mental illness must not be a barrier to the enjoyment of full social, political, and economic rights. The group tries to ensure that people are consulted before being deinstitutionalized.

In the United States advocacy groups use the press to advantage by focusing attention on welfare hotels and conditions in emergency shelters. Organizations like the Low Income Housing Information Service have research arms which lobby effectively to ensure that Congressional committees are aware of the extent of housing problems and the nature of difficulties encountered by people on the streets.

Education projects

In Britain several agencies devised training programs for homeless individuals preparing to move from shelters into conventional housing. The Housing Support Team started working in South London in 1981. It offers education to residents of hostels and staff training for service agencies to enable them to advise clients waiting to be rehoused. About 300 people are trained annually in a three-day course which covers social security procedures and benefits, budgeting and banking, housekeeping and establishing links with a new community.

The North London Education Project, initiated by the National Association for the Care and Resettlement of Offenders in 1980, supports education and job training for ex-prisoners in a residential setting. The center supplies shared accommodation and remedial tutoring, maintaining a formal link with the North London Further Education College. The project secures permanent housing for its graduates through local boroughs and housing associations.

Training programs

A number of projects in American cities are concerned with helping people become self-supporting. Project Self-Sufficiency in Los Angeles, which was organized by the Community Development Commission of Los Angeles County in partnership with a public–private task force, is directed at very low-income lone parents: over 25 per cent of families with children in Los Angeles are headed by a single parent. In parts of the county, the proportion of poverty-level families with children under the age of 6 is as high as 65 per cent. Project Self-Sufficiency offers personal and vocational counseling, life skills training, and career planning, while providing participants with child care, transportation, and mentors.

The Glasgow Homemakers' Scheme, funded by the city's housing department, gives training, counseling and support to homeless people moving into council housing or other flats. After initial instruction, follow-up home visits are arranged to ensure that new residents make the transition to permanent housing as smoothly as possible.

Several organizations in Britain offer training schemes to homeless people and to housing advisors and case workers in public and voluntary agencies who deal with homeless individuals. Shelter offers courses in Basic Housing Aid and Advice, Care in the Community, AIDS and HIV in Housing, Homelessness, Equal Opportunity and Housing, Immigration and Homelessness, Housing Finance, Building Inspections, Housing Law, Young People and Homelessness, and Women's Rights to Housing.

Health projects

Some individuals cannot contend with the welfare bureaucracies. Others are discouraged from using health clinics by long waits and hostile receptionists. They often are unsuccessful in obtaining proper care because they lack a fixed address or are known to be from hostels.

Several agencies in Britain now extend primary health care for people with no fixed abode. The East London Homeless Health Project, with government funding, encourages individuals to register with general practitioners. It attempts to involve medical professionals with hostels, HMOs, and B & Bs used by local authorities to accommodate homeless families. Progress has not been dramatic, however. Barriers erected both by homeless individuals and by members of the medical establishment are difficult to surmount. Health care is also being offered to immigrant communities in several London boroughs which now have "ethnic health services."

In Toronto nurses render medical assistance to street people. Known as Street Health (for men) and Street Haven (for women), these groups secured

46 Native woman receiving nursing assistance at Street Haven clinic in Toronto, 1988 (Photograph by Byron J. Bignell)

public funding to ensure the continuation of what were tenuous arrangements. Similar groups have been organized in Boston, Chicago, Philadelphia, Washington, and other American cities to furnish mobile health care teams which include a physician, clinical nurse, social worker, and mental health workers or a psychiatrist.

Operating in nineteen American cities since 1986, the National Health Care for the Homeless Program is predicated on a belief in the necessity of community-based health care. Program guidelines stipulate that proposals must come from a coalition including, at a minimum, homeless people, city and state governments, and local health care institutions. All projects must include at least a doctor, nurse, and social worker. Each project is required to arrange access to other services and benefits; for example, job finding, food, or housing services, and benefits available through public programs such as disability, workers' compensation, Medicaid, or food stamps. Demonstration projects are located in settings such as shelters, soup kitchens, missions, neighborhood centers, and detoxification facilities.

An assessment of this program by James Wright and Eleanor Weber determined that it is successful because of its community-based health care orientation, the dedication of workers, and the aggressive strategy adopted, including the use of former homeless people as outreach workers. The result, they concluded, is "a system of health care that is maximally accessible to a traditionally hard-to-reach segment of the urban poverty population" (Wright and Weber 1987: 154).

Substance abuse programs

The Hungerford Drug Project started in 1970 as a day center for drug users in London's West End. In 1978 this approach was abandoned in favor of counseling and outreach work, marking the first time that a specialist drug service in England had moved to the streets. Intended to maintain links with people who are unwilling to use conventional clinics, Hungerford offers telephone advice, an informal drop-in center, counseling, and outreach work among drug users. In a typical year project workers have 4,000 contacts with 1,000 different individuals. About two-thirds of the people who use clinic services are men; but 85 per cent of those who meet with the outreach workers are women; most are under age 20 and unable to claim benefits because they lack a local connection. Hungerford is associated with Turning Point, the largest national voluntary organization providing services for people with drug, alcohol, and psychiatric problems. Founded in 1964, Turning Point has 160 workers operating from fifty locations throughout England. They offer long-term "multi-substance misuse" residential rehabilitation programs, structured day programs, and counseling for those with alcohol or drug problems and their families, colleagues, or friends. Turning Point maintains self-contained accommodation to help people adjust to independent living after they have completed therapeutic counseling. Services were recently extended to various ethnic groups through a multi-lingual service in Southall, London.

AIDS programs

Street people are susceptible to AIDS. One of the problems encountered by people with HIV/AIDS is that they lose their housing (and often their employment) when it becomes known that they are ill. The AIDS Resource Center in New York City created two projects to maintain a supportive environment, permanent housing, and personal safety. Bailey House is a forty-four-person group residence offering comprehensive on-site services, ranging from nursing and medical care to recreation and counseling. The staff-to-resident ratio is 1:1, which allows for around-the-clock care. To gain admission individuals must be homeless, diagnosed as having AIDS, and eligible for income assistance.

A second initiative, the Supportive Housing Apartment Program, furnishes fourteen single and six family units on scattered sites. Services are tendered by a six-person mobile support team. Residents approve of the project because it approximates their pre-AIDS accommodations, promotes independence and a sense of community, while protecting confidentiality (Rose 1989: 64).

In a few cities AIDS clinics supply comprehensive services under one roof. A facility in Fort Myers, Florida, provides housing, food, clothing, medical care, and financial assistance. It operates an educational center which sponsors community seminars, meetings of AIDS support groups, and awareness programs for public school students. Operated by the Lee County AIDS Task Force, the clinic is funded by public and private agencies. Most of its financial support comes from area hospitals because the clinic enables AIDS patients to be diagnosed early, to receive appropriate care, and to significantly reduce the length of their hospital stay. Because many AIDS patients are indigent and incur high medical bills it is in the hospitals' interests to have them treated by the clinic rather than as in-patients.

Youth programs

Frequently young people are on the streets because they have no alternative. Some have been turned out by parents who no longer have room for them. Others are on the run from abusive domestic relationships or have recently left institutions or foster homes. Unless they have children or are mentally or physically ill, they do not qualify for public housing. As a result of involvement in drug trafficking and prostitution they are exposed to life-threatening situations. To assist them, such programs as Covenant House (in New York, Toronto, and other cities) operate mobile outreach services, drop-in centers, and dormitories for those under the age of 21. These projects refer teenagers to drug rehabilitation programs and assist them in locating job training and employment.

Outreach UK – Program for refugee youth

Outreach UK is an organization that works with unaccompanied refugee youth (from East Africa) to assist them in coping with life in Britain. Members of the Outreach team find private accommodation and obtain funding through government housing benefits. They also provide a rent guarantee scheme to landlords housing their clients, in the event of a default.

Mark Frere, Outreach team manager, describes their work as follows:

> the outreach team supports these young people in encouraging assimilation, education, health issues and cross cultural living . . . Also just being around . . . Outreach is street level social work and involves late hours coupled with many of the skills used by residential social workers. Knowing your way around the benefits system plus immigration law is a help . . . We manage the asylum applications [for these young refugees].

Outreach UK attempts to encourage independence in the youth that they assist. Frere notes that

> our experiences are showing that the refugees who arrive here are tough, resilient, and able to cope at the point of arrival. The patriarchal role played by local authorities erodes this independence and creates instead a condition of learned helplessness. We find this to be unhelpful and can lead to not coping well, failure and eventual homelessness.
>
> Education is their big drive and dream and offers another curious contrast. Young, serious, and diligent students totally committed to an education system that leaves them with very little, they will not consider working as an alternative "door of learning and integration." Result? 4 to 5 years on we are beginning to see young disillusioned men and women, unassimilated and with a deep sense of failure. Without work some are turning to drink and drugs and by doing so alienating themselves from their own community. Depression and all that goes with it follows. Suicide has been the choice of some . . .

An alarming rate of homelessness exists among young people. Because most distrust parole officers and other members of the establishment there is a need for advocates or mediators. The Handsworth Young Persons' Accommodation Committee began in 1981 by supplying interim housing for "young black offenders" in Birmingham's (U.K.) inner city wards. It extends counseling and advocacy services to ensure that young people leaving penal institutions have accommodation, health, welfare, and job training benefits.

A number of British initiatives work exclusively with young people. The Kaleidoscope Youth and Community Project, for example, operates an all-night drop-in center, a hostel and medical and educational facilities in southwest London for young people with drug problems. Some outreach organizations, funded (in accordance with the 1987 Children's Act) by local authorities in the center of London, work with unaccompanied young refugees from areas devastated by war or famine.

Programs for women

Often referred to as the hidden homeless, many women live in abusive situations or move frequently from one insecure accommodation to another. Single mothers, in particular, have limited choices on very small budgets. Because shelters and half-way houses for women are in short supply, women's groups (e.g. London's Homeless Action and Accommodation) have taken responsibility for housing in the absence of state initiatives. These organizations offer advice and referrals, provide counseling, and operate refuges for battered women and their children.

A few boroughs in London have publicly funded Women's Units to assist in securing housing, welfare benefits, and health care. In general, however, Britain lags behind North America in terms of the degree of sensitivity to women's issues and to the growing problem of homelessness among women. Many large cities in Canada and the United States have refuges or half-way houses for abused women as well as transitional housing that is integrated with community and social services (Daly 1990: 133-152).

Transitional housing for women and children

Groups like the Women's Institute for Housing and Economic Development in Boston developed transitional housing because women moving out of shelters could not locate affordable housing in the open market.

> Typically the residents of transitional housing are women who want to be self-supporting but need time, space, and help to reach that goal. They are generally required to enter into a contract with the housing sponsor/managers, a contract that delineates each resident's goals realistically within the program's time frame.
>
> (Sprague 1986)

47 Residents, volunteers, and staff of a women's shelter on a country outing, 1995 (Photograph courtesy of Joyce Brown)

Transitional housing is appropriate for women who are leaving an abusive situation. Many have never worked outside the home; they need assistance in planning their lives and preparing for a new job at the same time as they are trying to find permanent housing. Their children require support in making a difficult adjustment. This type of setting is effective for women leaving prison (70 per cent of whom are single mothers), for those recovering from drug or alcohol dependency, for people with disabilities, for widows and aging women, for teen parents, and for single parents who must relocate to a new community.

Some women are forced to leave home hurriedly, with only the clothes on their back. Thus, they need help to get re-established and most require services in addition to supportive housing. Included are child care and child development programs, life planning, job development, placement counseling, and referrals to health, welfare, employment, and community agencies. Some transitional homes also provide basic necessities like rental guarantees and furniture, as well as a revolving loan fund for security deposits which enables residents to find apartments on the open market (Sprague 1991). The key features of transitional homes are that they offer a secure environment, comprehensive support services and are predicated on the participation of residents in decisions affecting their future.

Supportive housing

Supportive housing, which is used extensively in Canada, generally is developed with government assistance by third sector sponsors. Housing is transitional or permanent. Supportive services are offered at arm's length or are available on the premises. In most cases, residents are involved in determining house rules and selecting new tenants.

An example is the fifteen-unit project sponsored by the Association for Women's Residential Facilities in Halifax with federal and provincial funding. Intended to provide independence and security for single women, the facility has 440-square-foot units, each of which contains an eat-in kitchen, living room, and separate bedroom as well as a deck and a small garden, along with a laundry and communal areas. A full-time staff person is available to residents for consultation. The building was completely occupied within two weeks after advertisements appeared. Sponsors feel that this arrangement is preferable to direct project development by government. They stress their understanding of the residents, many of whom they know as individuals. When planning a project they involve prospective residents: "there has to be both an element of choice and an element of participation. It must be a two-way conversation" (*Canadian Housing* 5, 5: 48).

In the past such voluntary efforts were criticized as being paternalistic. In the case of religious organizations participants were sometimes obliged to attend services or to "take the pledge." Sponsors now are more sensitive to these issues and many have stepped aside to let residents organize and manage themselves. They have also received criticism from communities for "dumping" group homes and half-way houses on unsuspecting neighbors. Because of this opposition and because many residents need and want permanent housing, most supportive housing consists of small, permanent projects or single homes which can be readily adapted to "fit" into existing neighborhoods.

Permanent housing for women and children

Some women need secure permanent rental housing which is convenient to public transportation, day care, employment centers, and health and social services. Several grassroots organizations have created their own housing. During the early 1980s the Women's Development Corporation in Providence, Rhode Island, designed and developed almost 300 rental housing units for low-income single mothers and children. The group carried out a participatory design process with prospective residents who helped to design, construct, and manage their own dwellings. The projects, which were the subject of a post-occupancy survey in the late 1980s, are considered successful by residents (Breitbart 1990: 18-24).

Programs for people with disabilities

Some people with disabilities are capable of living independently but cannot do so because services and appropriate housing are unavailable. They can be accommodated through adaptation of building design, by making dwellings fully accessible to wheelchairs, and by offering such essential services as domiciliary care. Frustrated by the obstructive attitudes of professionals and the patronizing views of charities, people with disabilities developed their own Centers for Independent Living in the United States (and later in Britain). These are local, non-residential agencies offering services under the direct control of people with disabilities. They provide information, counseling, suitably designed housing, appropriate technical aids, personal assistance, and transport.

Home-based support programs

Alternatives to institutionalization have evolved, based on the community care concept. The Centre François-Charon began operating apartments and group homes in Quebec City in 1979. It supplies accommodation and services in six locations for disabled people, ranging from hemiplegics to cerebral palsy patients. Most came directly from institutions. Support programs include 24-hour assistance for meal preparation, housework, and bathing. Social and escort services, extended for a four to six-month period, help to ease the transition to apartment living. Financial assistance is provided through a rent supplement scheme. Community integration programs are supported by a team consisting of a social worker, a nurse, a psychologist, and an occupational therapist.

Société d'Habitation du Québec, the provincial housing agency, played an important role in this initiative. It modified the selection criteria for subsidized housing to include people with disabilities; it financed the alterations to make housing accessible to wheelchairs; and it adjusted its standards so that all subsidized dwellings now meet accessibility requirements. After a decade of operation the Centre's programs cost about half as much as institutional care. Despite this encouraging experience, the Government of Quebec still spends sixteen times more on institutionalization than on home-based support programs (Plamondon 1985).

Cooperative housing and urban homesteading

Both the benefits and difficulties associated with cooperative housing and urban homesteading are illustrated by the case of Giroscope in northeast England. A group of young people, squatting and "living rough," got together to develop their own dwellings. All unemployed, they formed a workers' cooperative to purchase and renovate terraced Victorian buildings in the city of Hull. Each person has a private bedroom. All living areas are shared, as are maintenance chores and expenses. Workers in the cooperative do all renovation work, gutting the houses, rewiring them to current standards, damp-proofing, creating new bathrooms, replacing rotten woodwork and plaster, installing central heating and hot water systems, and rebuilding aged brickwork.

All members of the cooperative succeeded in housing themselves. Now, as each rehabilitated house nears completion they rent it out to low-income tenants (five per house) who complete the finishing touches. As these dwellings become available they are used as collateral for purchase of additional houses. At the same time as celebrating their successes, the members of the cooperative acknowledge that shared housing is a mixed blessing. Though the communal atmosphere is fragile, it allows people to share bills and forces them to learn to live more or less amicably with others. These skills would not be learned in bedsits, B & Bs, or hostels.

Giroscope's founders admit that their enterprise is on shaky financial ground. Average expenditures on each house are greater than operating revenues. They make up the difference with donations, grants, and deferred-payment loans. While pointing out that this effort is small in scale – and might not work in high-cost areas like London – they believe that the scheme can be replicated in British cities where derelict houses are available for under £12,000. Moreover, the Giroscope project creates employment for people on the dole, teaches building construction and office/book-keeping skills, regenerates vacant housing, and provides shelter for young homeless people.

Similar efforts, known as urban homesteading projects, are under way in American cities. These include rehabilitation of abandoned housing on New York's Lower East Side, creation of apartments by Chicago's Lakefront SRO Corporation, renovation of building shells by cooperatives in Boston, and re-development projects by the People's Homesteading Group in Baltimore where there are 5,800 empty houses.

Other rehabilitation and homesteading projects are being developed by large organizations like the Enterprise Foundation and the Local Initiatives Support Coalition in American cities. Founded in 1981, Enterprise forms partnerships with community groups to help them obtain low-cost housing, to secure support from local businesses, and to establish job placement agencies. The Local Initiatives Support Coalition (LISC) – started with funds from the Ford Foundation and major insurance and banking entities – operates in thirty-one metropolitan areas. In each city a local LISC raises funds; these are matched by the national umbrella organization and are used to finance residential, commercial, and industrial community renewal projects. LISC works through such groups as the Chicago Housing Partnership; funding is provided by LISC (start-up capital and low-interest bridge financing), private lenders (permanent financing), and the Chicago Housing Department (second mortgages). Labor is provided by neighborhood groups.

Shared housing

In order to preclude homelessness, especially among elderly people, formal home sharing arrangements evolved which allow individuals to split chores and living expenses. Shared housing ranges from self-initiated matches, to agency-assisted arrangements, to congregate or group homes where agencies are responsible for maintenance and management. Home sharing gives low-income people the opportunity to rent out rooms, and it affords older persons the option of receiving care within a private home. Some people provide services in exchange for board and room. This scheme permits elderly

48 Self-help housing organizations enable young people on the dole to learn building construction skills while renovating housing for their own use. Many are squatters or have been sleeping rough prior to moving into the rehabilitated dwellings (Photograph courtesy of Giroscope self-help housing group in Hull, England 1987)

individuals to remain in their own homes rather than being institutionalized. It increases the use of existing housing while reducing the need for public expenditures on housing and nursing home services. Shared housing offers single parents the possibility of receiving day care in exchange for other services.

Shared housing is not without its problems, however. Many potential participants have unrealistic expectations or are uneasy about sharing. The drop-out rate is high. Matches may be short-lived. It is difficult to find people willing to supply live-in services for elderly homeowners in poor health. Among older people shared housing may run counter to their traditional notions of privacy and independence.

Permanent housing

A number of schemes create permanent housing and security of tenure. These include rooming houses, short-life housing, and rehabilitation of existing owner-occupied units. Several organizations build SROs or rehabilitate existing structures. The Burnside Community Council in Portland, Oregon uses foundation grants to purchase and renovate SROs. Over a five-year period this group took control of 40 per cent of the downtown SRO and rooming house stock. They improved the units, lowered rents, and influenced other landlords to follow their example. After observing Portland's success Los Angeles funded SRO improvements of $19.2 million (Raubeson 1987: 101–102).

Another SRO was created in New York City when a non-profit housing corporation acquired a Skid Row hotel and obtained development funds from the State. The finished product includes forty permanent single rooms for single adults, twenty emergency units for homeless families with children, and a separate fifteen-unit wing for residents with psychiatric illnesses who require constant supervision. Residents share in the building's maintenance. Partial operating funds are provided by the New York State Department of Mental Health. The project was syndicated to secure permanent funding (Rosenbaum 1987: 102-103).

In San Diego, public financing, code modification, and a demolition moratorium were employed to preserve existing SROs and build new units. Regulations were revised to permit small living areas. This change, along with low-interest city loans, allowed developers to provide SROs with private bedrooms and shared living rooms and bathrooms in new or renovated buildings. Opposition was precluded by creating SROs in downtown commercial districts. Most of the units have been built by private developers. Rooms from 80 to 200 square feet rent for $300 to more than $600 monthly. Where public financing was provided the city stipulated that 20 per cent of the units, renting for $265 per month, must be set aside for low-income residents (*Planning* [1989] 55, 3: March).

In Toronto, a group of single homeless people and their advocates, organized as the Homes First Society, formed a non-profit housing corporation in mid-1983 to develop new SRO units. They constructed an eleven-storey "rooming house in the sky" in the city's downtown, consisting of seventy-seven

Enlisting the private sector

Andy Raubeson, Executive Director of the SRO Housing Corporation in Los Angeles, stresses the importance of eliciting public and private sector support:

> It is extremely important for anyone who endeavors to get involved in large-scale development to provide base-line housing for very low-income people to create the public will to support what you are doing. [Organizers must] include in that public will not just elected representatives but the movers and shakers of the corporate community.
>
> (Raubeson 1987: 102)

separate bedrooms. Every floor has two units, each with two or three bathrooms, one kitchen and living room, four or five bedrooms, and common areas. Each apartment is run as a separate household with its own ground rules. Every floor has a different key and is accessible only by the elevator from ground level. The project, which receives funding from the federal, provincial, and metropolitan governments, is linked to ten social service agencies that provide off-site support services to residents. As a result of a "deep subsidy" program rent is limited to 25 per cent of income; for most, this means that one-quarter of their welfare check will be used for rent. The project is run on the "facilitative management" model. Residents participated in the design of the building; 40 per cent of them now serve on building committees, and a number work as relief staff. They negotiate contractual agreements for house rules and resident selection. Bill Bosworth (1987: 84), former Executive Officer of Homes First Society, summed up this approach:

> If we focus on the fact that homeless people are severely debilitated, then we develop a service system in housing that continues the debilitation . . . Community groups are in the best position to put together the package of resources and programs that will create housing that can become homes.

Similar supportive housing projects were developed in Toronto for others who are considered " hard to house." In the mid-1980s All Saints Church and the Fred Victor Mission opted to close their emergency shelters and to provide a substantial amount of permanent accommodation in new, purpose-built structures for former street people. Jessie's, a refuge for pregnant teens and adolescent mothers with children, formed joint housing ventures with non-profit housing groups to establish housing cooperatives for low-income single mothers. Since 1977, Houselink has produced cooperative housing for people who have received psychiatric treatment. Residents share responsibility for decision-making, maintenance, and planning.

Housing planned and managed by users

Homeless people are accustomed to having others make decisions for them. It is difficult for potential users who are homeless to assist in the design of their own housing: many are transient, some have problems getting along with others, and a number lack experience working in groups. Moreover, government programs which require quick results and extensive paperwork tend to discourage participation.

Nevertheless, experience demonstrates that many homeless people can make the transition to permanent housing if they have a voice in the planning and operation of their own dwellings. It is necessary to have committed, full-time community organizers to recruit people from shelters, soup kitchens, and from groups such as those comprised of people with disabilities. Professionals, including architects and planners, must be educated so that they do not impose design solutions on users. Government officials must allow sufficient time and flexibility for this process to evolve. It is important to have continuity in public programs; when requirements change frequently those who are not experts in the field are excluded.

In spite of these obstacles, a number of projects were developed which involved potential users in planning, design, and management. The Inuit Non-Profit Housing Corporation, which deals with Canada's Native population, has included users in the housing design process since the late 1970s. The initial projects took two years to build but have prospered because users are a part of the process and have a stake in the maintenance of their housing.

The Native Council of Canada worked with Canada Mortgage and Housing Corporation to provide 1,000 homes, under the Urban Native Housing Program, for Natives who left reserves to find jobs in cities. Eighty non-profit housing corporations were formed by Native groups to work with CMHC in developing housing for low-income families and seniors. Natives hold the deeds and manage the projects (Lang-Runtz and Ahem 1987: 44-46).

In Vancouver's Skid Row, the Downtown Eastside Residents' Association (DERA) was formed in 1973 to preserve the community and its housing. Ninety per cent of the people are on fixed incomes; 80 per cent are men, more than half are disabled, and their average age is 57. Because there was no legislation in British Columbia protecting roomers these individuals could be evicted without notice or cause, rents could be raised by any amount, and their belongings confiscated by landlords, as happened during the construction of Expo '86, when 750 pensioners in the DERA district were summarily evicted.

Members of DERA decided to create self-contained, permanent housing in order to protect their tenancy rights. Jim Green, the group's Executive Director, said that "it is essential that the people who are going to live in the housing are members of the design team and participate directly in creation of projects." Green noted that this process helps individuals to develop their own skills in dealing with the world:

> What we do is create architecture of opportunity, giving people the opportunity to alter their lives through housing. To design and operate their own homes in an atmosphere of security helps people develop pride and an awareness of their own abilities. It teaches them how to deal with people and how to handle defeat.
>
> (*Canadian Housing* 5, 5: 50)

Preservation of existing housing

Short-life housing offers the possibility of long leases for people without homes. This mechanism, widely used in Britain where there are about 750,000 empty dwellings, involves the use of properties which have been borrowed from their owners for periods of 1-5 years. Typically owners are local authorities and housing associations or large corporations like British Rail. Properties are licensed or leased to a cooperative or housing association, which is responsible for repairs and management. Occupants, most of whom are homeless, pay minimal rents to the lessee. Many organizations specialize in sheltering such groups as women, minorities, and people who have left institutional care. The residents are able to occupy housing which would otherwise stand vacant while awaiting demolition or rehabilitation. Though not permanent, short-life accommodations offer some security of tenure, particularly because leases are often renewed when owners see that properties are being well maintained or improved.

An adequate supply of permanent housing can be assured by maintaining or enhancing the existing stock. Care and Repair Ltd. operates projects in Britain designed to assist elderly home-owners who are in danger of becoming homeless when their dwellings fall into disrepair. Building repairs, which the owners cannot undertake on their own, are paid through loans arranged with financial institutions and housing grants from local authorities. The process assures that older people, at virtually no cost to the public treasury, can remain in their homes, rather than being institutionalized. One of the Care and Repair programs, known as Staying Put, is offered in thirty-four locations by the Anchor Housing Trust; help has been given to 25,000 people over the past decade. Half of the clients are over age 75; over half are singles; almost three-quarters have health problems; half are on low incomes; and about 40 per cent live in pre-1919 properties which require major maintenance.

Sheltered housing

Anchor is also a major provider of sheltered housing. This form of accommodation, usually for older people, is common in Britain. Anchor Trust, for example, houses more than 24,000 tenants in more than 21,000 sheltered units. A typical project has about thirty units, often in a single-storey configuration with communal facilities, including a central kitchen and dining area, emergency alarms, and a resident warden. The project is about evenly divided between one-bedroom units and bedsits (studios). Approximately three-quarters of residents are women; about two-thirds of all tenants have poverty-level incomes (less than £80 per week in 1992) and rely on state benefits.

Self-help housing

The final category is self-help housing. It is given a great deal of attention because it offers considerable potential for homeless people who have the desire and capacity to live independently. Residents have built their own housing in a number of cooperative ventures. These include homes in the

49 Poor conditions in homes owned by pensioners motivated Shelter to start Care and Repair Ltd. to assist older people in obtaining assistance to effect repairs to their dwellings (Courtesy of Shelter)

Problems with self-build

Redlining
In Tucson we had very good advice from a real-estate man . . . walk through the old neighborhoods looking for hand-written "For Sale" signs. Since the old barrios are all red-lined (I know it is illegal but it is true) there are no bank loans/no mortgages/no house insurance . . .

Time and costs
So we bought a two-bedroom, eighty-year old adobe house. It took me two years to rebuild it and pay off the seller's loan. I spent three months studying construction, talking to contractors, etc. . . . it is not possible to build a house here for less than $40,000: $20,000 materials and $20,000 labor. Now if you can do much of the work yourself . . . you can get near the $20,000 materials cost but not under it, as used materials are not approved for construction unless each piece is certified by an engineer.

Incremental building not permitted
But the major obstacle that the codes put up is the illegality of incremental building . . . I cannot build a one-room shack to live here, save money, work, add a bathroom, enclose the outdoor kitchen/oven, add another room for the new children (etc. . . .) as I get time/money/skills. Yet this sketches how my current house was built from 1919 to 1947.

The banks/financial institutions along with construction institutions have used government police powers to halt all such incremental building. You must pull a permit for all work to be done, and complete the work within a year. No occupancy permits until finished.

(Richard Williams,
e-mail communication, July 11, 1995)

inner London borough of Lambeth, rural schemes in Canada, and rehabilitation projects in American cities. Some of the house builders received intensive training in the trades and went on to become successful contractors. In addition to sweat equity or voluntary assistance, self-help initiatives rely on funding from a variety of sources. These range from government grants, subsidies, and tax incentives, to corporate and foundation matching funds, to contributions of land and technical assistance from public or private sources.

The community architecture movement in England includes a number of self-help projects where young people on the dole pooled their resources to construct or renovate dwellings for their own use. Following its success in Lambeth, the Walter Segal Trust organized a national campaign to replicate these projects, to set up practical building courses across the country, to provide technical support, and to secure financing commitments from building societies and housing associations.

Rural projects are developed in the United States by very low income families in Appalachia and by Natives on tribal lands. The Kentucky Mountain Housing Development Corporation (KMHDC) is a joint venture of church groups, government agencies, and the private sector. Working in rural areas of Appalachia since 1973, it deals with unemployment, homelessness, and poor housing, using sweat equity by member households, and financing provided

50 The buildings rehabilitated by self-help groups in Britain must be completely gutted before renovation work can commence. Most of these structures pre-date World War I and require entirely new or re-cycled plumbing, electrical and heating systems
(Photograph courtesy of Giroscope self-help housing group in Hull, England 1987)

51 Community architecture project, Black Road, Macclesfield, England, 1986. This movement, which is responsible for a number of creative rehabilitation projects in several British cities, represents an attractive alternative to wholesale urban renewal (Photograph courtesy of Rod Hackney)

by the U.S. Farmer's Home Administration and the KMHDC Home Loan Program. The average loan for KMHDC is only about half of the average loan for the principal government programs used in the region.

Established as a private non-profit organization in 1971, the Utah Navajo Development Council (UNDC) constructs homes for very low-income Navajo on their reservation. The UNDC encourages self-reliance by offering educational programs and vocational training for children and adults, along with health clinics and housing. In a year's time the UNDC completes more than thirty new homes. No land cost is incurred as the dwellings are built on tribal lands. Homes are 960 square feet, with two or three bedrooms. Virtually all of the funding for these homes comes from UNDC, which earns revenue from its mineral resources and the sale of livestock. The tribal council effects savings by operating a job training program, enabling families to participate in the construction of their own homes after they have completed courses offered by the tribe.

A variety of self-help schemes in the United States are sponsored by private groups like the Enterprise Foundation which has developed 6,000 low-income housing units. These projects provided construction jobs for 10,000 low-income residents. They range from new low-cost housing for rural residents of Maine to the rehabilitation of ghetto housing in such places as Chattanooga and Cleveland.

A number of self-help strategies were devised by tenants to improve conditions in multi-family buildings as a way of precluding homelessness: one of these was the renovation of a 270-room SRO in Pittsburgh. There is considerable scope for creative operations because inner cities have hundreds of acres of abandoned housing. In North Philadelphia, for instance, about 10,000 units stand empty (Kolodny 1986: 447-462). Another project, to convert buildings abandoned by landlords to resident-run cooperatives, was successfully completed by low-income people in Harlem, many of them older women (Leavitt and Saegert 1989).

In Canada self-help housing production accounts for almost 15 per cent of all new construction and more than 40 per cent of residential renovation. In the Maritime provinces (Nova Scotia, New Brunswick, Prince Edward Island, Newfoundland, and Labrador) half of all residential construction and more than 60 per cent of single-family dwellings are built by self-help labor. Some of the self-help initiatives are described here.

Cape Breton Labourers Development Company started a revolving house construction fund to finance new homes for its members. Homeowners' payments are returned to the fund to finance more homes, once construction is complete. Homes are not elaborate, which makes it easier for people, with help from semi-skilled friends and relatives, to build their own dwellings. Typically, they are one-storey bungalows, built in standard sizes (20 × 28 feet, for example) to take advantage of standard 4 × 8 foot building materials (dry wall, exterior siding, and interior wall board are 4 × 8 feet; studs are pre-cut to accommodate an 8-foot wall). But they do meet user needs and expectations, they satisfy local building codes, and they are consistent with community construction standards.

Small homes were built by extremely poor farm workers in Nova Scotia, many of whom previously lived in buses and shanties. A grassroots effort, Hearth Homes, began with workshops on budgeting, maintenance, and building methods. Eventually fifty-three families constructed their own dwellings in standard 22 × 26 foot modules. Financial assistance came from the provincial government and employment training benefits were provided by the federal government. Families rent these homes for $350 per month, with payments credited toward purchase.

Street City is a housing project initiated by street people in Toronto who rehabilitated a former post office warehouse, starting in the winter of 1988/89. Funding of $30,000 was supplied by the City of Toronto and a capital grant of $600,000 was obtained from the Ontario Ministry of Community Services. Housing seventy-two formerly homeless men and women, the center has small individual rooms (8 × 12 feet) opening on to an interior street. Each room has a bed, chair, dresser, and a small refrigerator. The residents organized a conflict mediation committee, run their own store and banking service, and share maintenance duties. Several of these projects are described in Table 13 which indicates the nature of the organizing group, the residents, costs and financing methods, and the type of construction or housing provided.

Table 13 Examples of self-help housing in Canada

	Organizing group	*Users/ participants*	*Housing type*	*Costs/financing mechanisms*
Cape Breton Labourers Development Company	Labor union	Union members	Self-built single family	Revolving fund using payroll deductions; $350 monthly payment
Downtown Eastside Residents Association (DERA)	DERA	Very low-income unattached singles from the neighborhood	Rehabilitation of single room occupancy units and old hotels	Financing by various government agencies; residents' welfare payments defray project costs
Hearth Homes	Community-based non-profit group	Very low-income farmworkers	Simple, woodframe, modular houses	Government financing: user cost $350/ monthly; rent counts toward home ownership
Homes First	Private non-profit housing corporation	Very low-income people from shelters	New and rehabilitated apartment buildings with single units and shared common areas	Financial contribution by government; residents' welfare payments defray project costs
Habitat for Humanity	Habitat volunteers	Low-income households	Single family	Donations and sweat equity; cost is $350/month
Government Rural and Native Housing	Government agencies and Native groups	Primarily Natives in remote areas	Single family; self-help construction with government technical assistance	Government financing; forgivable 5-year mortgage
Street City	Street people	Homeless singles	Renovated warehouse	Municipal/ provincial grants

Despite the popularity of self-help, particularly in Canada's rural areas and in the Maritimes, it is generally not a viable alternative for the neediest (at least in the major cities) because of the high cost of urban land and materials, and the impediments posed by building and zoning codes, as well as the restrictive attitudes of lenders and government officials. Government assistance is required on such matters as financing, technical support, guidance in dealing with building codes, and site acquisition.

The benefits of self-help include provision of housing at reasonable cost, community economic development, fostering other community initiatives, and the development of community cohesiveness. Moreover, because of sweat equity self-help builders generally do not incur much debt financing, which aids them in weathering economic downturns. It is a process which allows people to construct over a long period of time, adding on as finances permit. Many build in their spare time, while employed full time, using skilled friends and relatives for specialized tasks, often on a barter basis.

Table 14 summarizes some of the characteristics or attributes of self-help housing. It emphasizes the elements of user control over space, control over the design and planning processes, as well as the concept of access in social, spatial, economic and political terms. This matrix stresses the importance of self-help in fostering individual and community identity and sense of place. It suggests that the social and spatial dimensions of self-help housing are widely accepted in principle; but the economic and political dimensions, shown in the bottom half of Table 14, may involve considerable difficulty in implementation because of the need to change existing practices in financing and development regulation. This may necessitate shifts in existing practices or traditional thinking not only by bureaucrats and lenders but by consumers as well (Daly 1993a: 20-27).

Self-help is time consuming and requires considerable determination, planning and management expertise, as well as patience. Some individuals who initiate the process have no building skills and, without proper supervision, they may produce less than satisfactory housing. There are also barriers with respect to acquiring affordable, properly serviced land, obtaining construction financing, and securing necessary approvals. Among the problems cited by self-builders are changes in economic conditions, financing terms, government programs and the regulatory climate, as well as opposition by existing home owners who wish to exclude low-cost housing. In addition, institutional lenders and government officials often are wary of the perceived risks inherent in self-help housing. A related issue is development standards; often these are set nationally without regard to local conditions, materials, and practices.

Self-help projects are easier to produce in rural areas where land is cheap, most people have known their neighbors for some time, many residents have building skills and are accustomed to bartering, and where government regulations and enforcement are minimal. In cities, conversely, self-help initiatives face serious hurdles and are subjected to greater scrutiny by government bureaucrats. Moreover, the sites for urban projects may be less than ideal; a few local authorities in England and the United States have

Table 14 Attributes and characteristics of self-help housing

	Control	*Access*	*Identity*
Social	User control over design of interior and exterior space. Able to live where they want, with whom they want	Access may be facilitated by government agencies or community organizations	Bottom–up planning and decision-making enhances sense of pride and identification with housing and with neighborhood; enhanced community spirit
Spatial	Incremental building allows individual to start small and maintain control over the development process and the design; opportunity to test low-cost and efficient house designs	Space may be designed and built to suit family's particular requirements and desires. Community self-help groups can ensure access to communal spaces	User identity and sense of ownership increased by ability to shape the residential environment. Sense of place a key factor
Economic	Can build incrementally as finances permit; increased security of tenure as self-help may make home ownership possible. Reduced need for government control and financing during and after construction	Cost-effective means of providing affordable housing; lower financing requirements, reduced front-end costs, and lower mortgage and carrying costs enhance accessibility, affordability, and availability	Self-help households often develop housing with others in a like-minded community; possible long-term community economic benefits. Identity may be enhanced by job development, economic spin offs, and skills training/apprenticeship
Political	Bureaucrats may be reluctant to relinquish control to users. Regulatory bodies are still able to exercise control over the building and planning processes	Access to surplus government land not yet a reality for self-help housing groups. Some government agencies assist by providing financial and technical aid	Maximum involvement by users; self-help is sometimes used in concert with other political consciousness-raising activities

turned over derelict housing estates for renovation by local groups. Harloe, Marcuse, and Smith note that these opportunities may present a two-edged sword: "They can genuinely tap residents' powers of collective organization and action, but may also be seen as an invitation for the marginal poor to manage their own confinement to a sort of inner city reservation" (Harloe *et al.* 1992: 198).

Alternatives to homelessness

Advocacy/litigation
- Publicize conditions and problems
- Legal aid to preclude homelessness
- Litigation to ensure that people receive entitlements
- Pressure public agencies to fulfil responsibilities

Information
- Information, housing advice, and advocacy centers located (e.g. at welfare offices) where potential clients gather

Health
- Street-front health clinics
- Drug outreach projects/counseling
- AIDS programs, including hostels and supportive housing
- Home-based support in lieu of institutionalization

Education and training
- Links between education, training, and housing
- Direct connection between training and actual jobs
- Life skills, budgeting, and related courses
- Linkage of job training with welfare benefits, health care, transportation, and day care

Housing
- Non-profit and cooperative projects, including joint ventures between public, private, and voluntary agencies
- Retention of SROs; creation of SRO apartments
- Residential intensification
- Urban homesteading
- Short-life housing
- Community architecture and planning
- Shared housing
- Accessible housing for people with physical disabilities
- Alternative zoning practices which permit development of social housing, group homes, and SROs
- "Care and Repair" grants for low-income home-owners
- Self-help: housing planned and managed by users

SUMMARY

The American experience demonstrates that important initiatives can come about through joint public–voluntary endeavors. Foundations play a decisive role in the American effort to deal with homelessness. Both Canadians and Americans develop self-help housing and small-scale transitional and cooperative housing for women. Several projects are effective in involving residents in

52 Prince Charles congratulates members of the Rebuild self-build group in Edinburgh, October 1989. *Roof* reported that, in 1986, 11,200 homes were constructed by self-builders in Britain (Courtesy of Shelter)

project planning and operation by means of facilitative management. The development of social housing in Canada and non-profit projects in the United States demonstrates the advantages of joint public–private ventures where the major responsibility for shelter provision rests with private non-profit and cooperative entities, representing groups of low-income and moderate-income residents.

British programs, emanating primarily from the voluntary sector, but funded in part by government, confirm the importance of education, lobbying, and advocacy. A number of projects successfully integrate housing, employment or training, and support services. Those dealing with vulnerable groups like ex-prisoners and low-income single mothers affirm that anticipatory planning can preclude homelessness.

The direction taken in each country is somewhat different. There is substantial variation in terms of the nature and extent of participant involvement, the roles of public and voluntary agencies, and the strategies employed to preclude or minimize homelessness. This review points out the possible advantages of examining one another's experiences in dealing with this complex issue. It also demonstrates the central importance of understanding the historical, cultural, and institutional context in which programs have evolved, of appreciating what is unique and what is generic within the policy environment. It does not follow, therefore, that it is possible or desirable to attempt wholesale cross-national transfer of programs or projects. It is preferable to think in terms of adaptation and modification to suit local circumstances; to stimulate creative projects by allowing participants to develop new ways (suitable to the local situation) of addressing known problems after evaluating the experience of others.

12

RECOMMENDATIONS

In the late 1980s and early 1990s global economic changes led to high levels of poverty, rising unemployment in certain cities or regions or among particular groups, and to a widespread shortage of affordable housing. Any proposed solutions, then, must deal with these issues. In each of the three countries problems vary from region to region, from city to city, and from neighborhood to neighborhood. Accordingly, locally devised, community-based programs are most appropriate. Yet, because homelessness is also a nationwide problem, the federal (or central) government must be involved and must provide sufficient funding on a continuing basis to assure permanent remedies. This is a necessary element in our social contract. Despite the validity of the concept of irreducibly social goods, however, it would be naive to ignore the climate of fiscal restraint in the 1990s. Government expends public funds on questionable defense systems and on bailing out savings and loan companies. But it seems unlikely that major social spending initiatives can be expected. Moreover, "our collective view of what government can and should do for its people is being continually narrowed" by the contented majority who challenge the need for public assistance to the disenfranchised (Galbraith 1992: 30).

These recommendations, then, are premised on the efficacy of self-help and predicated on reducing expenditures on stopgap measures. Public spending, instead, should be directed to self-help initiatives and non-profit groups that can produce permanent housing by using such devices as tax incentives, mortgage-backed securities, and joint ventures with developers.

The experience of New York City, which in the early 1990s was spending $250 million annually on emergency shelters and welfare hotels, illustrates the futility of "temporary" responses. As soon as cots are added, more shelterless individuals join the queue. Institutional responses of this nature, which seem to exacerbate chronicity, succeed only in patronizing homeless people, emphasizing their disabilities and social isolation, while maintaining them in a cycle from which escape is difficult. Instead of reserving public funds for housing and related social services, significant amounts are paid to the owners of welfare hotels and to professional care-givers, some of whom have a vested interest in the continuation of homelessness.

Traditional responses to homelessness require critical rethinking. Governments and certain charitable organizations have been inclined to do things *for* homeless people. This asylum mentality views individuals on the street as pathologically ill, deficient, and incompetent. Services developed from this perspective are devised as a social prosthesis, a crutch to prop up the weak (Sullivan 1992: 206). But many of those who are, or have been, homeless stress that they are a resource unto themselves. They want to be involved in building their own shelter and in shaping their own futures. These individuals and groups are best equipped, in most cases, to determine what they need in terms of housing and community services and where these facilities should be located. Moreover, programs must not be imposed from above but must involve the users in design, planning, and implementation. The concept of cooperative or self-help housing should be exploited as it represents a rare congruity among homeless people, liberal advocacy groups, and conservative governments.

Consideration must be given to the working poor because they are at risk and because the provision of work incentives will reap long-term rewards. Work has real and symbolic importance, particularly for individuals below the poverty line. Among the strategies suggested are raising the minimum wage to its former levels (where it provided an incentive to leave welfare), offering adequate medical coverage, providing credits to partially offset the cost of daycare, instituting refundable tax credits to raise the level of take-home pay, and increasing the value of earned income credits. Job creation may range from using inner city residents to improve the infrastructure in their neighborhoods, to practical training for service and technical jobs. At the same time, not everyone can find work in the new economic order created by globalization. Some, like mothers with young children, must not be compelled to work but should be encouraged to continue to stay with their children (Jencks 1994: 110).

We must move away from the emphasis on "special needs groups." This approach has two unfortunate consequences. First, public officials tend to limit programs and funding to those with particular "needs," such as people with physical or mental disabilities. Second, when criteria for admission to housing and related programs are predicated on illness, alcoholism, or drug dependency, individuals tend to adopt the behavior patterns necessary to get them admitted.

Contrary to prevailing stereotypes, those who are homeless are distinguished mainly by their poverty and their heterogeneity. Included are families, single mothers and children, visible minorities, runaway youths, and those suffering from long-term unemployment. Some, like battered women, are hidden from public view, vulnerable, and often ignored. Accordingly, a range of programs and services is required. Not a proliferation of services which reproduces the asylum but services which buttress individual and community networks and reinforce their capacity for self-help. Beyond the requirement for permanent housing, these programs should include advocacy and information centers (in public places like neighborhood libraries); education; job training and counseling; and a variety of health schemes and outreach projects.

Programs must be achievable and pragmatic. The most successful efforts have been small in scale, tailored to community needs, with limited, realistic

53 Former homeless resident, completing electrical work for Mace Housing Association, England (Photograph by Brenda Prince, courtesy of Shelter)

objectives. Self-help housing, for instance, will never be a huge undertaking and it is not without limitations. With public funding (redirected from current emergency shelter spending) and private sector assistance, however, it is possible for this type of initiative to be more broadly applied. It is likely to be most suitable for rural areas; but a related form of self-help, renovation of existing housing – sometimes with partial funding from the public sector – has been employed to create decent dwellings for low-income residents of inner cities.

Permanent, affordable housing must be a basic building block of any proposed solution. Dwellings must also be livable and secure in terms of privacy and tenure, with some degree of tenant management. This may require assistance from all levels of government. The issues of housing size, scale, physical design, and management require attention. Government involvement often means large, bureaucratized projects; these will not work. Small clusters are needed, as in the case of self-managed group homes, where the residents set house rules, screen new tenants, and determine their own direction. The question of participation in design and planning decisions is especially important for those with disabilities and others who have been marginalized by current "shelterization" policies (Timmer *et al.* 1994: 184).

There must be significant federal (or central) government involvement on a continuing basis in funding demonstration schemes, in subsidizing operating projects and programs, and in research – particularly into means of preventing homelessness. Provincial or state governments must assume a more active role in creating housing and in providing support and funding for municipal and non-profit programs. Local governments should participate in devising and implementing local public programs and joint public–private projects. All units of government can help by providing vacant or under-utilized public land and buildings for housing. Local authorities can assist the process of providing affordable dwellings by instituting "one-stop application processing" and taking other steps to break processing log-jams. Municipalities can relax excessive building codes and standards to encourage the development of SROs and to allow for rehabilitation of existing housing stock. Many of the housing standards embedded in codes are inappropriate, unnecessary, do not meet user needs, and result in buildings which are far too expensive. Average house size has ballooned, especially in North America since the end of World War II, when many privately built homes for the families of returning veterans were less than 800 square feet in floor area. Affordable housing must be truly affordable; to make it so requires reducing expectations. It is instructive to examine the self-built dwellings in Canada's Maritimes: they are basic, small by current standards, but acceptable and affordable to their owner-builders.

A related concern is the preservation of affordable rentals and the protection of tenant rights. The application of "just cause" eviction ordinances, for example, can be required by local authorities to safeguard tenants. In addition, temporary assistance should be provided to renters and home-owners experiencing crises, in order to avoid eviction. Municipal governments have an obligation to preserve SROs, rooming houses, and other low-cost housing. Local governments should ensure, in cooperation with the voluntary sector, that group homes and other appropriate accommodations are provided for those now forced to use emergency shelters.

Governments cannot be expected to shoulder full responsibility. Considerable scope is available for the third sector. Non-profit groups and non-government organizations have extended their activities to include the development and management of permanent housing alternatives. Voluntary sector involvement is crucial because public housing and conventional market housing generally are not available to those who are potentially homeless. The experience of Canadian non-profits and numerous groups in the United States demonstrates that non-profit development and management (with some public assistance) represents a feasible alternative to public construction of housing projects. Public agencies also should induce the participation of the private sector. More attention can be given, by public and private agencies, to innovative joint programs which facilitate development of affordable housing.

There is no panacea to the issue of homelessness; it is a complex problem which will not disappear overnight. Nevertheless, even complex problems can be addressed, perhaps in a variety of ways and by a range of actors. Agencies involved in the provision and funding of housing and social services must make a long-term commitment to alleviate problems associated with homelessness. Though tempting to public officials, progress should not be measured by the extent to which city streets can be freed of "derelicts"; nor is the number of emergency shelter beds an adequate measure of compassionate reform. Instead, the housing and services provided should be judged by these standards:

- *affordability* for the poorest fifth of society
- *appropriateness* of the social services and housing provided to suit the needs of homeless individuals and households
- *accessibility* to public services, to health care and social services, and to the courts to ensure rights and to combat discrimination
- *availability* of housing in areas where it is most needed
- *effectiveness*: support services should be adequate to meet needs but they must not induce dependency; they should maximize the individual's ability to function independently
- *power* and user control, as reflected in resident participation in decisions regarding the design, cost, location, and management of housing and related community services
- *production*: increased output of new dwellings, major rehabilitation, and repair and maintenance to make better use of existing stock and to ensure adequate shelter with respect to housing type, cost, location, and dwelling characteristics
- *equity*: fairer access to and a more equal distribution of housing programs and social services across cities or regions
- *security* for people and their belongings; not just physical protection but also peace of mind, freedom from excessive regulation, and from violence or intrusion
- *prevention*: measures designed to preclude homelessness.

13

COMPARATIVE OBSERVATIONS AND CONCLUSIONS

In this chapter I will comment on some of the similarities and differences among the three countries. I will attempt to link these conclusions with the conceptual framework set out in the first chapter by addressing the cross-national context of homelessness, the different notions of state and society in the three countries, the effects of global economic shifts, their historical trajectories in evolving social and housing policies, and the complementarity of individual and community roles. I will argue that different responses to homelessness are conditioned by our disparate notions of individual, community, society, and state; these concepts, in turn, are products of our culture and history.

CROSS-NATIONAL COMPARISON OF RESPONSES TO HOMELESSNESS

Lipset asserts that "nations can be understood only in comparative perspective" (Lipset 1990: xiii). Charles Haar concluded that cross-national comparisons may prompt questions about what is frequently taken for granted and may lead to "insightful reflection" (Haar 1984: 1). The three countries in this study are linked by language, culture, politics, and economic/trade relations. They share many commonalities in their social history; most American and Canadian social welfare institutions as well as the public health apparatus evolved from the British paradigm. But there are important differences among the three in terms of cultural environment and organizing principles which help to explain variations in their responses to homelessness. Moreover, the brief review of Western European experience (in Chapter 4) provides a stark reminder of the links among homelessness, immigration, employment, race, and poverty, and of their devastating impact on destitute newcomers. In virtually all of the countries examined, social benefits to the poor have been eroded in recent years, along with regressive shifts in tax burdens, marginalization of those in the bottom income quintile, and growing disparities between the haves and the have-nots.

I believe that it is possible to develop a better understanding of responses to homeless people by comparing these countries in terms of their definitions of the state and civil society, the individual and the community. Britain's response, for example, is based on its long history of collectivism, a considerable faith in the efficacy of the welfare state, a general deference to authority, and an emphasis on group rights. British housing policy embodies notions of class and gender which were meant to benefit solid working-class voters and preserve the nuclear family. American social policy is still shaped to a considerable extent by considerations of race, especially with respect to blacks. Lipset notes that "slavery and racism have been the foremost deviation from the American creed" (Lipset 1990: 182). The clouds of slavery still hang over urban ghettos in the United States, continue to have an effect on race relations, and are conspicuous in the country's bifurcated social policies. Canada never tolerated the institution of slavery. Its population, until recent times, was quite homogeneous, as it was favored by immigrants from the British Isles. Its social policies have not been as controversial as those in the United States, at least until the recent past when it became more common to question authority. Canada's policies aim to equitably allocate public goods, to ensure that particular groups like the elderly and prospective home-owners are looked after, and to promote job creation.

The British experience illustrates the necessity of a social safety net, the need for a public role in housing creation, and the importance of a vital voluntary sector. Despite a variety of innovative programs and projects, the record in Britain underscores the necessity for government participation. It demonstrates that housing provision alone will not suffice. Without public financing of housing, economic development, and training programs many who receive assistance will remain on the dole without hope of finding employment and of helping themselves. Social policy cannot be divorced from employment policy; what subverted the welfare state's aim of achieving greater social equality was high unemployment. That is why some critics find workfare such a limited concept. It is essential to create meaningful jobs before people can be put to work, if the program is to have the desired long-term results (Townsend 1993).

The recent evolution of British policy on homelessness is closely tied to ideology and politics. It is meant to reflect the ruling Conservatives' emphasis on preservation of family values, to enlarge the ranks of home-owners who are likely to vote Tory, to give more scope for the operation of market mechanisms, and to lessen the power of local authorities in terms of direct housing provision. Britain's social spending, though low by European standards, is higher than Canada's, which in turn is higher than the United States'. As a consequence of British redistributive policies those in the bottom quintile are significantly better off after transfer payments. Though inequalities have become more pronounced in recent years, the British system does not contain disparities as large as those between the rich and poor in North America. Despite its flaws, the homelessness legislation in Britain has been on the books and has been enforced to a significant degree across the country, benefiting millions of people who have been housed under the Act since 1977. By identifying priority groups and insisting on a local authority duty to house people immediately on being found homeless, the Act sets a useful precedent for government action to deal with those who are at risk.

Britain and Canada are more supportive of group rights and communitarianism while Americans favor individualism. Canadians think of themselves as unlike Americans in the sense that they are more reserved, civil and deferential, and less achievement-oriented or competitive (Lipset 1990: 55). The notion of competitive individualism is at the heart of the American response to homelessness. The United States' founders associated the state with oppression; they distrusted rulers and believed in self-reliance, egalitarianism, and populism, reinforced by a tradition of voluntarism (Lipset 1990: 2). Traditionally, American charities differentiated between deserving and undeserving poor, while emphasizing social control and the need for assimilation and conformity. These strands reappear in today's social welfare policies in general and in responses to homelessness in particular. Welfare rules are punitive and the system, despite its claims, operates as a disincentive to work. Homeless people who are forced to the street by a natural disaster are well looked after because they are thought to be deserving; but those who fail to conform to social norms represent the irksome other, are shunted aside, and often find they cannot receive their entitlements.

The American constitution, written after a revolution, ensured that power was dispersed, that checks and balances were in place to restrict government's scope, particularly to limit its ability to intrude into individual and local affairs. The United States is a nation-state held together as a federation. Each state and municipality can, to a significant extent, determine its own destiny. Social activists find this particularly troubling when the federal government shirks responsibility and when the byzantine nature of local government thwarts innovation. American local government is characterized by autonomy, fragmentation, and a very limited notion of the public good; these characteristics are manifest in parochialism, a tradition of patronage and corruption, and a narrowly construed view of the public interest.

Although Americans believe that their country offers equality of opportunity, it is accepted that there will always be haves and have-nots. It is a society built on polarities; perhaps this is to be expected, given the country's history, immense size, and its extremes . . . geographically, economically, and socially.

The roots of today's homelessness in the United States, then, lie deep in American history. Housing policy traditionally favored the middle class, and officials stigmatized people who were homeless or dependent on public assistance. Aversion to charity and to government intervention is widespread. Given the widely held beliefs in individualism, laissez-faire capitalism, private sector solutions to social problems, the efficacy of social control, and a conviction that government should not be an innovative force for change, current attitudes toward homeless individuals in the United States are predictable. Today, these notions are exhibited in a stultifying array of rules governing welfare recipients, policies extolling the virtues of self-reliance and home ownership, and in a belief that everyone must be rated on a scale of economic usefulness.

Canada is influenced both by the British model and by the immediate, overwhelming presence of its neighbor to the south.The Canadian governmental system is relatively streamlined. The federal government was given considerable powers because the population was relatively small and dispersed, centralized power was essential to develop the country's resources, and a strong national government was seen as important to keep American expansionism in check. In recent years, much of this power has been transferred to the provinces.The Canadian local government system (based on the British model) has a great deal of uniformity, operates within a single legal framework, and is subject to effective administrative control by a higher level of government (Cullingworth 1993: 194). It is not fragmented, generally does not operate along party lines, and exists mainly to deliver municipal services. In recent years provinces and cities have become much more concerned with policy, in terms of economic development, growth control, regional distribution of services, urban–suburban conflicts over taxes and services, transportation, housing issues and homelessness.

Canadian social policy is beneficial to those in the bottom income quintile; in the recent past their income share tripled after taking into account the effect of transfer payments and income taxes. Housing policy, in its formative years, was meant to help those who were relatively well-off members of the working class; recently, it has been reasonably successful in promoting income mix in public, not-for-profit and private (but publicly subsidized) projects. These policies have helped to prevent homelessness, to preserve the vitality of urban centers, and to avoid the rush to suburbia which characterized post-war settlement patterns in the United States. In the mid-1990s, however, retrenchment at the federal and provincial levels is a cause for concern. The federal government is backing away from earlier commitments in the housing and social welfare areas, while several provincial governments are moving to the right, reducing social expenditures, and casting covetous glances southward, surveying the privatization scene in the United States.

The Canadian situation demonstrates the importance of differentiating among regions because of substantial variations in economic health, the crucial necessity of a social safety net, the importance of a national health service which is free to low-income people, and the desirability of assisting people before they lose their homes.

In Britain, Canada, and the United States central government has devolved most authority for dealing with homelessness to the local level, sometimes with adequate funding to do the job, sometimes not. This is generally acknowledged as a good thing, because the government closest to the people is thought to be the most efficient and responsive to local concerns. Often this is not the case. In the United States one out of eight workers is now employed by government; well over two-thirds (11.1 million people) are state and local public employees. Total employment at the state and local levels has been increasing much faster than at the federal level. This has not resulted in efficiency. On the contrary, state and local governments, as in the case of the savings and loan scandal, are easily seduced by lobbyists for special interests. In many cities municipal government is a hodge-podge of overlapping jurisdictions, each fiercely protective of its turf. Recent experience demonstrates unequivocally that devolution is not a panacea. Without oversight from higher levels of government and from the voluntary sector, many local officials would ignore homelessness.

GLOBALIZATION

Along with devolution, neo-conservative governments in all three countries during the 1980s and 1990s reacted to global economic changes by deregulating and privatizing public sector activities. Social services, in particular, have been marked for reduction and contracting out to the private sector. When government opts out of these areas for-profit private enterprises fill the void, often paying minimal wages and offering little security to a marginalized work force of ethnic minorities and recent arrivals, many of whom are women. The presence of these groups in urban centers has accentuated class and racial polarization and new forms of spatial disparity as the middle class flees to protected communities on the city's fringe. The areas they leave behind, marked by poverty and a concentration of non-whites, are seen as peripheral and problems associated with the inner city, like homelessness, are ignored. Poverty and homelessness intersect, with their effects concentrated on identifiable groups: single mothers, children, and non-whites are disproportionately affected. In racially polarized American cities, the abandonment of the collective is indisputable as the middle classes vacate the center, opting for the gated suburb. This exodus may be seen as a social parallel to the political and economic movements of privatization and deregulation. Because they worsen disparities among income groups and are predicated on a reconstruction of Social Darwinism, these processes yield a relatively narrow range of perceived solutions to the issues of poverty and homelessness.

LANGUAGE AND ACTION

The relationship between language and action is particularly important to a study of homelessness. Our approach to the study of homelessness is predicated on ideology, or structures of ideas shared across a society or culture, which construct our world view. These ideas, from each culture's prevailing ideology, are captured by our language, are taken as received wisdom and become embedded in policy (Livingston 1994: 177).

An example of this phenomenon is housing policy, an area in which the mode of political discourse is determined and shaped to fit an ideological mold. By setting the agenda interest groups are able to frame the debates and to dictate the pace and direction of change. An illustration of this is the way homelessness has been defined by governments and pressure groups. In Britain the legislation sought to strengthen people's rights to housing; some local authorities circumscribed these rights in practice; subsequent media attention put pressure on government to broaden the scope of its ameliorative actions. This is a classic case of "power and conflict over the social construction of a housing problem" in which the legislation was created by advocacy groups, brought forward by a private member, passed after the Department of Environment was brought on board, then undermined by local authorities in court actions, and strengthened again in the courts as appeals of arbitrary local actions were brought by homeless people (Kemeny 1992: 31).

Thus, a nation's definition of homelessness frames its response. In the United States a narrow definition is used, at least by public agencies, and the numbers thought to be homeless are low relative to Canada and Britain. Local authorities in Britain are obliged by law to house people without adequate shelter; as a result, the numbers certified as homeless are relatively high and the nature of those accommodated is different. British law specifies that older people, pregnant women, psychiatrically ill individuals, and some singles must be considered vulnerable. Homeless people in American cities, on the other hand, are primarily singles and young mothers with children. Canada is located somewhere between the American and British approaches. The customary use of the term 'homelessness' in Canada includes some people (particularly battered women and youths) who are considered to be at risk.

The fragmented nature of the American federal system, division of powers, and multiplicity of agencies all work to the detriment of a broad definition of the public good and have undermined welfare reform initiatives. The programs in place to assist homeless people are narrowly construed and thus are vulnerable to cutbacks. Stopgap reforms designed to address a crisis like homelessness are likely candidates for retrenchment (Slessarev 1988: 377). Reform is unlikely in the United States unless it duplicates successful social programs of the past,

> those that combined bureaucratic capacity for innovation and legislative maneuvering with congressional support, in order to deliver benefits directly to a wide array of citizens regardless of their private incomes and without demeaning "means tests" or other cumbersome application procedures.
>
> (Weir *et al.* 1988: 437)

Liberals and conservatives agree that the American welfare system needs to be overhauled. It discourages initiative and self-sufficiency. Its anachronistic and stifling culture is intractable and damaging to the low-income people who are compelled to submit to its demeaning rituals: "The current system discourages clients from working not only through its financial incentives but also through its bureaucratic impediments [which foster] an eligibility-compliance culture" (Bane and Ellwood 1994).

Social policy in the United States has been two-pronged since the adoption of the Social Security Act in 1935: Social Security for the middle class and respectable workers; welfare for the irksome others, typically women of color with young children. The political isolation of the poor, especially the black poor, from the working and middle classes was "a bitter legacy of the Great Society . . . a deepening of the rifts within the ranks of actual and potential supporters of U.S. public social provision" (Orloff 1993: 309). Middle-class whites favor such programs as Social Security, which is based on the notion of beneficiaries earning their rewards; but they are reluctant to pay taxes for other social programs that benefit the poor, because these are construed as "handouts." Differentiations between the deserving and undeserving remain with us.

THE ROLE OF THE INDIVIDUAL AND THE CONCEPT OF COMMUNITY

Underlying assumptions of the role of the individual and the concept of community help to explain the differences in the ways we view homeless people: as the deserving poor, deviants, victims, or social misfits. I suggested in the first chapter that homelessness may be the result of a loss of sense of community as a consequence of such forces as globalization. I agree with Charles Taylor that there are irreducibly social goods and that one of the principal functions of the state is to ensure that these are distributed equitably. But the power and the will of the state are limited, particularly in those societies that define issues such as homelessness, at least in part, as beyond the state's mandate. When the lack of shelter is seen primarily as a personal problem or evidence of a personality defect, it is relatively easy to dismiss both the issue and the people involved as superfluous. We define the world as a self–other dualism, relegating the other to the periphery of our consciousness. Our societies, and particularly the large, global cities, are distinguished by atomistic behavior and an extraordinarily high number of one-person households. Many of us deal with complex social issues by denial or reductionistic reasoning, epitomized by an assumption that homelessness can be dismissed from our consciousness after we give to the United Way and donate a can of tuna to the food bank.

The individual self is dominant in our thinking and this preoccupation occludes a sense of community. "Communities . . . are blessed to the extent that they are aggregations of blessed individual souls, not as deserving entities in their own right" (Livingston 1994:104). The notion of community is central to civil society; it refers to a sense of belonging, a climate of participation, user control, mutualism, interdependence, cooperation, reciprocity, and group consciousness. In this view of the world the individual and community are not mutually exclusive; their spheres overlap. I am suggesting that we begin to see people without shelter as integral to group consciousness and, therefore, politically and socially capable of participating in programs which they help to design and manage.This strikes me as the most likely path out of homelessness for them and out of complicity in the construction of homelessness for those who are not without shelter.

What I have attempted to show in this study is that homeless people are a dynamic, heterogeneous group with capabilities that are overlooked or submerged by professional care-givers. Robert Coles describes them as:

> a whole potpourri of individuals, some of whom . . . are sane and solid and need some money. Some of them have some money but not enough to pay for rent in a city that just doesn't have available housing for them. Some of them are out-and-out alcoholics and psychotics, or down-and-out people who in the old days used to be in the Bowery or in places . . . where they could find some kind of community. And now we've gentrified these places.
>
> (Coles quoted in Giamo and Grunberg 1992: 186)

Many of these people have considerable coping skills and, from a more creative perspective, could be viewed as a resource rather than as a drain on society. Individuals receiving welfare assistance generally cite their desire to "get off welfare" as one of their top priorities; and most do – the great majority of welfare recipients in the United States are off the rolls in the first four years. The desire among homeless people to get off the streets is equally strong. Elliot Liebow's interviews with women in shelters reveal that most, if they qualify for housing vouchers, are able to obtain decent housing. Once their housing situation is secure they are much better equipped to stabilize their lives (Liebow 1993: 237–249).

If allowed to exercise power and control over the programs which shape their futures, many can become advocates for themselves. A number already have. To realize this ambition, however, professional care-givers must relinquish some of their control. "Case management" approaches must give way to self-help and self-advocacy initiatives. Line workers and managers in social service agencies are obliged to shift their focus away from infantilizing and patronizing their clients to something more akin to the peer counseling approach used by self-help groups. Homeless individuals themselves must be allowed to regain their individuality and self-esteem and to assume considerable personal responsibility for following one of the paths out of homelessness. Initially, though, even self-help efforts will require some funding assistance from the public sector and perhaps space and support from the voluntary sector:

> What else is there, if it's not personal responsibility? And who am I to take it away as a desirable thing from anyone? And who am I to condescend and patronize people by saying they are totally helpless, and the only thing that will get them by is me and my kind and our programs? Personal responsibility is our last chance in this life.
>
> (Robert Coles, quoted in Giamo and Grunberg 1992: 199)

It seems appropriate to close with the words of a formerly homeless man in Oregon who observed ruefully that "too often people experiencing homelessness and poverty are given programs rather than given the opportunity to design the programs" (John Statler, October 3, 1995). To move in the preferred direction of group consciousness in a civil society new approaches, including a variety of self-help initiatives, are needed. I hope that this comparative study of social policies and strategies will help us to start along that path.

APPENDICES

APPENDIX A

Chronology of housing initiatives in Britain, the United States, and Canada

	Britain	*United States*	*Canada*
1850–1899	Lodging housing legislation passed (1851)	New York City police furnished lodging to 435,000 per year (1875)	Immigration at mid-century and at end of century. First settlement houses established in late 1800s Toronto
	Torrens Act regulated public works and slum clearance (1868)	Philanthropic capitalists, model home builders, settlement houses, and Charity Organization Society active (1880s)	
	Formation of Local Government Board (1871)		
	Housing of the Working Classes Act; government involved in housing (1890)		
1900–1919	Local authorities allowed to build council housing outside towns (1900)	New York City Tenement Housing Department formed (1901)	American tenement laws a model for Toronto (1910)
	Purchase of Letchworth site by First Garden City Ltd (1903)	Boston imposed height restrictions on downtown tenements (1904)	Royal Commission reports on wartime problems, including poor housing (1919)

cont.

	Britain	United States	Canada
	Hampstead Garden Suburb designed by Parker and Unwin (1907)	Massachusetts Homestead Commission formed to build homes (1910)	
	London County Council completed its 8,000th house (1909)	Zoning introduced in New York City (1916)	
	Increase of Rent and Mortgage Interest (War Restrictions Act) (1915)	Defense housing produced by two federal agencies (1918)	
	Housing and Town Planning (Addison) Act; Ministry of Health formed (1919)		
1920–1929	Housing Act (Chamberlain subsidy, 1923)	President's Conference on Home Building and Home Ownership (1931)	Homelessness and severe poverty evident in cities
	Wheatley Housing Act; Prevention of Eviction Act (1924)	U.S. Home Loan Bank Act (1932)	
	Greenwood (Slum Clearance) Housing Act (1930)		
1930–1939	Housing Act provided subsidy to relieve overcrowding (1935)	Public Works Administration (PWA) Housing Division created (1933)	Relief camps established for single homeless men (1934)
	Housing (Financial Provisions) Act (1938)	Federal Housing Administration started (1934)	The Dominion Housing Act provided $20 million in loans to finance 4,900 housing units over three years (1935)
	Imposition of full rent controls (1939)	First public housing and greenbelt towns (1935)	The Federal Home Improvement Plan provided interest rate subsidies on rehabilitation loans (1937)

	Britain	*United States*	*Canada*
		U.S. Housing Authority formed; PWA completed 22,000 units (1937)	The National Housing Act offered assistance to home buyers and provided for the modernization of existing housing (1938)
1940–1949	4 million houses completed by government in inter-war years (1940)	Lanham Act for defense housing (1940	The Wartime Housing Corporation (1941) was absorbed by the Central Mortgage and Housing Corporation, later the Canada Mortgage and Housing Corporation (CMHC) (1946)
	Ministry of Town and Country Planning established (1942)	Veterans' Emergency Housing Program; GI Bill passed (1945)	First joint federal–provincial public housing programs (1949)
	New Towns Act (1946)	Housing Act authorized slum clearance and redevelopment (1949)	
	Housing and Town Planning (Green Belt) Act (1947)		
	The year of the Welfare State; National Assistance Act passed (1948)		
	Parker Morris housing manual (Homes for Today and Tomorrow) (1949)		
	Public housing no longer restricted to "the working classes" (1949)		
1950–1959	Ministry of Housing and Local Government formed (1951)	Landmark legislation in urban renewal (1954)	The Bank Act was amended to allow chartered banks to lend mortgage money (1954)

	Britain	United States	Canada
	Slum clearance restarted; private sector improvement encouraged (1954)		
	Subsidy structure encouraged high-rise housing (1956)		
	House Purchase and Housing Act; public loans to building societies (1959)		
1960–1969	Government given compulsory powers for district urban renewal (1964)	Community Mental Health Act recommends deinstitutionalization (1963)	The federal government allowed loan transfers up to 90% to the provinces to build provincially owned public housing (1964)
	DHSS commissioned Greve study on homelessness (1967)	War on Poverty started by President Lyndon Johnson (1964)	The federal Rent Supplement Program paid the difference between market rent and 25% of household income for low-income households in private rentals (1969)
		U.S. Department of Housing and Urban Development created (1965)	
		Civil Rights Act (Title VIII) establishes fair housing as national objective (1968)	
1970–1979	Fair rents introduced for council tenants (1972)	Nixon imposed moratorium on new construction of public housing (1973)	The National Housing Act was amended to provide financial assistance for new home buyers, loans for cooperative housing, and low-interest loans for municipal and private non-profit housing; The National Housing

	Britain	United States	Canada
			Act encouraged the integration of different income groups in social housing projects (1973)
		Housing and Community Development Act promoted block funding approach rather than categorical programs (1974)	The Rural and Native Housing Program provided new dwellings and renovation assistance for low-income Native and non-Native people in rural areas (1974)
1980–1989	Housing (Homeless Persons) Act (1977)	Government encouraged private investment in low-cost rentals (1981)	
	Introduced statutory right to buy for tenants in council estates (1980)	National Coalition for the Homeless established (1982)	
	Councils required to help people vulnerable to homelessness (1985)	President's Commission on Housing recommended more reliance on private market (1984)	The federal government redirected social housing programs to households in greatest need; housing program delivery was devolved to the provinces and territories (1986)
	Legislation broadened the definition of homelessness (1987)	HUD report on emergency shelters (1984)	Major international conference held in Ottawa to mark the International Year of Shelter for the Homeless (1987)
	Government induced landlords back into private rental market (1987)	Stewart B. McKinney Homeless Assistance Act passed (1987)	The down-payment required for homes purchased with the backing of CMHC was reduced from 10% to 5% (1989)
	1.5 million council homes sold during the 1980s (1989)	Suspension of discretionary contract awards due to HUD scandal (1989)	

	Britain	*United States*	*Canada*
1990–1995	Single Homelessness (Rough Sleepers) Initiative in London (1990)	National Affordable Housing Act established HOME housing grants (1990)	The federal government budget reflected a reduction of $51 million over two years in funds for low-cost housing (1990)
	Empty Homes Agency established to accommodate homeless households (1992)	Low Income Housing Preservation and Resident Home Ownership Act (1992)	The federal government terminated the cooperative housing program which had created more than 60,000 homes for low- and moderate-income households (1992)
		Funds reserved for community-based non-profit housing developers (1992)	
		HUD introduced "continuum-of-care" approach (1993)	
		Gingrich's "Contract with America" (1995)	

APPENDIX B

Parallels and links in the evolution of housing and social policy in Britain and North America

<table>
<tr><th>Britain</th><th>United States and Canada</th></tr>
<tr><td>In the mid-1800s pauperism in Britain led to pressure for deportation and emigration</td><td>The resulting immigration into former colonies in the United States and Canada helped to fill jobs created by industrialization</td></tr>
<tr><td>Starting in the 1860s, "5 per cent philanthropists" built model homes in England for better-off working people ("artisans")</td><td>Later, model dwellings were built for "honest, hard-working mechanics" in American cities</td></tr>
<tr><td colspan="2">The Salvation Army started in Britain and was soon transplanted to North America</td></tr>
<tr><td>In 1880 the Charity Organisation Society was founded in England</td><td>It was then replicated in North American cities; its work was carried out by "Friendly Visitors" who sought to promote social order and to acculturate the "deserving poor" among the immigrants</td></tr>
<tr><td>In 1884 the settlement house movement began in London</td><td>By the end of the century there were over 400 settlement houses in the U.S. and Canada</td></tr>
<tr><td colspan="2">England had a history of philanthropy and reform, much esteemed by middle- and upper-class Americans, who made annual treks to London to learn what was being done to cope with urban problems, ranging from sanitation and public health to housing. Repeated epidemics, particularly in the slums, and the discovery of germ theory, led to the introduction of sanitary reform and the public health movement. Sanitarians from North America began their work by visiting with colleagues in London's slums</td></tr>
<tr><td>Housing legislation in England (1890) was intended to ease overcrowding and to encourage building of quality accommodation; the Act heralded a transformation of the state's role, from negative to positive intervention, and marked the influence of reformers arguing the merits of gradualist socialism; by 1909 the London County Council housed 46,000 people</td><td>In the United States and Canada, however, tenement legislation (1901) was narrow and restrictive, resulting in the closure of unsafe buildings, and making no provision for new housing</td></tr>
<tr><td>Britain was a more homogeneous society in which emigration to the colonies served as a safety valve</td><td>Because North American cities were growing so rapidly and there was a huge influx of foreigners, public discourse related to issues of social control, good government, assimilation, and ways to deal with the "dangerous classes"</td></tr>
</table>

cont.

Britain	*United States and Canada*
Because of a longer exposure to urban problems, an awareness developed earlier in Britain of the connections between major social and economic issues: poverty, low wages, unemployment, the casual nature of jobs, and the immobility of the poor were inextricably linked with overcrowding, ill health, high rents, and housing shortages in the cities	
World War I resulted in such hardships that Britain felt compelled to build "homes for heroes"	The U.S. and Canada began their first national public housing programs, based directly on the British model
After the end of World War I, concerned about rising class-based discontent, Britain offered new housing. Nevertheless, industrial disruptions were widespread, culminating in the General Strike (1926). The state began the inter-war housing program which created 4.1 million homes by 1939, 37% of which were subsidized	In the U.S. and Canada the post-war era was marked by labor unrest, deportations, the Red Scare of 1919, and immigrant restriction. Joblessness and strikes were commonplace and wage rates often were below the minimum income needed to support a family. But neither the American nor the Canadian government intervened. Housing and social services were left to the voluntary sector
By the end of the 1930s many working-class Britons were better housed, though the poorest third of the population were unable to qualify for council housing. Location decisions in England generally were made at the local level; many authorities preferred to curry favor with their upper working-class or middle-class constituents by constructing new suburban developments, which created improved housing, jobs, and a source of votes, while inner city problems were ignored	"Housers" in North America were strongly influenced by British and European experiments: Frederic Howe, *European Cities* (1913); Edith Elmer Wood, *Housing Progress in Western Europe* (1923); Catherine Bauer, *Modern Housing* (1934). The 1934 "Housing Program for the United States" was adopted by housers with European experience; they were assisted by Raymond Unwin of Letchworth Garden City. Their program, which contained numerous references to recent housing developments in Britain, became the basis for the U.S. Housing Act of 1937. The public housing program that evolved in the U.S. was substantially influenced by the British experience. By the end of the 1930s, however, most working-class Americans and Canadians remained both poor and poorly housed
	Early city planning efforts in the U.S. and Canada were modelled after English Garden Cities; this was evident in World War I defense housing (1918), Sunnyside Gardens (1925), Radburn

cont.

Britain	*United States and Canada*
	(1927), and the greenbelt towns (1935). British town planners, including Raymond Unwin and Thomas Adams, worked in Canada and the U.S. on regional plans and new towns
New Deal reforms of the 1930s were echoed in the Beveridge Report (1943) and the formation of Britain's welfare state	In Canada the Marsh (1943) and Curtis (1944) reports relied on Beveridge and pointed to European experience in recommending universal social security and cooperative housing
At the end of World War II all political parties in Britain endorsed broadening the scope of council housing. Tower blocks were built in areas where housing had been demolished during the war. Housing was linked to social services as an integral part of the welfare state	In the United States and Canada post-war housing served to create construction jobs, to relieve overcrowding, and to stimulate "urban renewal" with an emphasis on monolithic public housing projects and large urban commercial or civic projects. Most new residential building in the U.S. was in the suburbs, spurred by the availability of 90 per cent federal subsidies for highways and by the G.I. Bill, which allowed veterans to purchase homes at favorable rates
Council housing in Britain housed the better-off members of the working class. It was used by politicians of all political parties to secure votes, to generate goodwill, and to provide tangible evidence of progress toward the goal of decent housing for all	Public housing was, to a large extent, seen as a job creation and social control device and was most readily accepted in times of disequilibrium as a temporary response to social crises. In the post-war years public housing became the shelter of last resort in American cities, housing an increasing percentage of low-income single mothers and ethnic minorities
Later, as system-built tower blocks in inner cities deteriorated, units were occupied by those who had no reasonable alternative. By the mid-1990s a very substantial minority of tenants had no earned income; many were classified as homeless by local authorities	In Canada, because of a shift to cooperative and mixed-income social housing, these problems, while still evident, have not been nearly as severe
British legislation with respect to housing and homelessness is more centralized. Central government sets policy and controls the purse strings, using regulations and the *Code of Guidance* to rein in local authorities. But implementation is still largely a local function and there are considerable disparities between housing *policy*, reflecting ideology and legislative intent, and housing *programs* carried out on the	In general, housing and homelessness are more important functions of municipal government (in Britain) than in the United States and Canada, though large Canadian cities have taken a major role in the provision of social housing in recent years

cont.

Britain	*United States and Canada*
ground. The relationship between central and local governments is dynamic, shaped by shifting political winds as well as by the vagaries of the economy	The U.S. and Canada have federal systems with considerable power exercised by the states and provinces. Tension between national and local desires is evident, but local forces, guided by parochialism and home rule, often dominate. It is difficult to coerce localities to build housing if they are not motivated. As a result, local efforts are often exclusionary, prompted by economic concerns, and tend to reinforce existing racial/settlement patterns
By the mid-1990s Britain had sold off 1.5 million council homes and was moving toward a form of privatization, relying heavily on housing associations	The United States has opted out of public housing, has sold off public projects or turned them over for tenant management, and has promoted privatization and joint public–private schemes; Canada drastically scaled back the role of government in housing provision and, because of the high costs of mixed-income social housing, devolved more responsibility to localities, the private sector, and voluntary groups

NOTES

U.S. dollar billions are U.S. billions (a thousand millions); pound sterling billions are a million millions.

List of magazines/ newsletters referred to in text

Roof – published by Shelter, London
Rumours – a Roomers and Boarders Association Newsletter, Toronto
Employment Gazette – published by the Department of the Environment, London
Housing Associations Weekly – published by National Federation of Housing Associations, London
Fact File – the statistical bulletin of the Housing Corporation, London
Now Magazine, Toronto
Chronicle of Higher Education – published by Editorial Projects for Education, Washington, D.C.
Canadian Housing – Canadian Association of Housing and Renewal Officials, Ottawa
Planning – American Society of Planning Officials, Chicago
Globe and Mail, Toronto
The Observer, London
Toronto Star, Toronto
HUD User – U.S. Department of Housing and Urban Development, Office of Policy Development and Research
Guardian Weekly, London
New York Times Magazine, New York

Photographs sourced as 'U.S. Archives' are from the Public Housing Administration historical files, U.S. National Archives and Records Service, Still Photographs Branch, Washington, D.C.

1 INTRODUCTION

1 Subsequent references generally will be to Britain (i.e. England, Wales, and Scotland). When necessary, however, citations will be for England only, for England and Wales combined, or for Scotland, because statistics are gathered in a variety of forms for these areas by different agencies. Ireland refers to the Republic of Ireland and does not include Ulster.
2 In 1994 the low-income cut-off for a four-person household in Canadian cities was $31,256 in Canadian dollars (equivalent to $22,500 in U.S. dollars); the American poverty line for a non-farm family of four was $15,141.
3 "Collectivism" refers to the treatment of social questions by the state with a view

towards effecting a more equitable distribution of resources than is possible under a system of laissez-faire capitalism. It emerged in reaction to individualism and its inveterate distrust of the state. In emphasizing the priority of the community rather than of the individual, it asserted that certain social problems were beyond the scope of private initiative, that ameliorative measures required the extension of government regulation over the economy, and that society's wealth could be redistributed through welfare programs.

4 Standards relating to minimum acceptable dwelling size, number of rooms, and amenities (hot water, heat, toilet, etc.) are defined by each country's housing agency and are used to collect census data and official statistics.

2 POVERTY AND THE POLITICS OF DISTRIBUTION IN BRITAIN, THE UNITED STATES, AND CANADA

1 Housing associations build houses which are financed by a 75 per cent cash grant – reduced to 67 per cent in April, 1993 – from the Housing Corporation. The remainder is borrowed from a bank or building society. The tenant's rent covers the interest and repayments on the private loan as well as the housing association's fee for management and maintenance. Between 1988 and the end of 1992 private sector lenders provided £2.4 billion as well as another £1 billion to purchase local authority housing.

2 A full-time worker, earning $5.30 per hour, would make about $10,600 per year. In 1990 the U.S. government defined the poverty level for a family of three as $10,419.

3 Proponents argue that a subsidy is necessary to promote ownership. Yet Canada, without such a subsidy, has the same rate of ownership as the United States.

4 Americans who receive health care insurance from their employers are permitted to exclude the value of this benefit from income for tax purposes.

4 IMMIGRATION AND HOMELESSNESS IN EUROPE

1 An earlier draft of this chapter appeared in G. Daly (1996) "Migrants and gatekeepers: the links between immigration and homelessness in Western Europe," *Cities* 13, 1: 11–23.

5 A PRESCRIPTION FOR POOR HEALTH

1 Earlier drafts of this chapter appeared in three articles. G. Daly (1989) "Homelessness and health: views and responses in Canada, the United Kingdom and the United States," *Health Promotion Journal* 4, 2: 115–128; G. Daly (1989) "Homelessness and health: a comparison of British, Canadian and US cities," *Cities* 6, 1, Feb.; G. Daly (1990) "Health implications of homelessness: reports from three countries," *Journal of Sociology and Social Welfare* 17, 1: 111–125, March.

7 THE EXPERIENCE OF HOMELESS PEOPLE

1 According to American definitions, housing affordability problems exist when renters pay 30 per cent or more of their income in rent, or owners pay more than 35 per cent of their income for mortgages and maintenance, or 30 per cent and higher for maintenance of a mortgage-free dwelling. Crowded housing is defined as units in which there is more than one person per room. Physical adequacy relates to plumbing, electrical and heating systems and the presence of kitchen and bathroom facilities.

2 These figures are from "The Politics of homelessness," including statements by Jim Green, Executive Director, Downtown Eastside Residents' Association, Vancouver, in Lang-Runtz and Ahem (1987: 63).

3 The suicide rate among young people in the Yukon and Northwest Territories (over 50 per 100,000 people) is five times higher than the national average. In the Western Arctic a survey discovered that eight out of ten Native girls under the age of 8 are victims of sexual abuse, as are 50 per cent of Native boys the same age (Mary Beth Levan, study co-sponsored by the Northwest Territories Women's Association and the Government of the Northwest Territories Social Services Department, Yellowknife, N.W.T., January 28, 1989).

8 MORE THAN JUST A ROOF

1 New council dwellings can be built for less money than is paid for B & Bs or for the refurbishment of a typical local authority dwelling. In 1994 the average daily cost per household placed in B & Bs was £40 in London and £26 outside the city. A major stumbling block to new construction in cities like London, however, is the difficulty of acquiring reasonably priced land in districts where housing is needed.

2 This situation was attributable in part to the federal government's Emergency Assistance Program. Though families on AFDC (Aid for Families with Dependent Children) are limited to about $300 monthly for shelter, the government spends up to ten times as much to house them in welfare hotels if they are classified as homeless. Hotel operators are allowed to charge whatever they like. The City of New York pays more than four times as much per room as do universities which negotiated contracts for similar hotel rooms to be used by students. Non-profit groups like Henry Street Settlement House provide medical care, social services, and housing for homeless families at a cost 20 per cent less than that incurred by the municipality for housing alone.

3 At Seaton House, like many hostels, the initial event is an interview with an intake worker who tries to ascertain whether the applicant has exhausted all other avenues for shelter and care. The worker will limit the length of stay (usually ranging from three days to two weeks). If shelter is still needed at the end of that time the individual must go to other hostels and return to Seaton House with proof that the other hostels are full.

9 FROM SHELTERS TO PERMANENT HOUSING: THE EVOLVING ROLE OF GOVERNMENT

1 By 1990 London boroughs were spending about £150 million annually on temporary lodgings; the annual cost per household, £14,600 in London, was substantially more than the annual cost of building a new home to rent (£8,200 in London). There were 28,500 families in temporary accommodation in London in 1991; by mid-1992 the total had increased to 44,600 families.

2 In fiscal year 1987 the McKinney authorization of $442.7 million was reduced to $355 million. For fiscal year 1988, the $616 million authorization was cut to $357.6 million; for fiscal year 1989, the $634 million authorization was reduced to an appropriation of $387.4 million; for fiscal year 1990, the authorization of $676 million was reduced to $600 million. The 1991 budget submission cut low-income programs by more than $2 billion, after adjusting for inflation. All but four McKinney Act programs were reduced in real terms. The administration also sought to eliminate the Low Income Weatherization Program and the Low Income Housing Energy Assistance Program.

3 The National Affordable Housing Act of 1990 established housing grants under the HOME program; this was strengthened in 1992 with the passage of the Housing and Community Development Act which reserved 15 per cent of the

funds for community-based non-profit developers. The Low Income Housing Preservation and Resident Home Ownership Act of 1990 sought to prevent private owners from opting out of projects which would have resulted in displacement of low-income tenants. But it took HUD more than one year to develop regulations for implementation of this legislation.

4 The SSI program helps very poor elderly and disabled persons. In 1991 people with no other income received a monthly payment of $407 (individuals) or $610 (couples) from the federal government. The states may add to this if they choose. Only half of the states offer supplements and most have frozen benefits for at least the past five years. General assistance is a state and local program of last resort to help those low-income people who do not qualify for AFDC or SSI. Only 60 per cent of the states offer this assistance. The monthly payment averaged $215 per person in 1991, representing 39 per cent of the poverty line. States also offer Medicaid which provides health care coverage for poor families with children, low-income elderly persons, and people with disabilities. Recipients of AFDC and most of those receiving SSI are entitled to Medicaid but those on general assistance are not. This program is jointly funded, paid for in a 58:42 ratio by the federal and state governments.

   Poor people in the United States are eligible for food stamps and some receive supplemental assistance. The Special Supplemental Food Program for Women, Infants and Children provides food, counseling, and nutritional guidance to pregnant women and those with children under the age of 5. To receive benefits their family income must be below 18.5 per cent of the poverty line and they must be at nutritional risk. Only 55 to 60 per cent of eligible families receive these entitlement benefits.

5 In Washington, D.C., for instance, the number of AFDC recipients increased by 20 per cent in just three years, 1991–94. In the eight years prior to 1995 the number of city residents with jobs declined from 320,000 to 250,000. One in eight households was without a telephone. More than four out of ten black men (ages 18–35) are in jail, on probation or parole, or sought on arrest warrants.

6 Affordable housing is defined by Ontario as "housing which is affordable to households of low and moderate income." One half must be affordable to households with incomes up to the 30th percentile and one half must be affordable to households with incomes between the 30th and 60th percentiles of incomes for the region. Income data for each region of the province are based on census returns which are then escalated by the Consumer Price Index. "Affordability means annual housing costs (gross rent, or mortgage principal and interest [amortized over 25 years and assuming a 25 per cent down payment], and taxes) do not exceed 30 per cent of gross annual household income" (Ontario Ministry of Housing 1989).

7 In Canada social housing is produced, with public funding, by public non-profit housing corporations, by housing cooperatives, and by private non-profit housing groups. As the term is used in Canada, social housing refers to publicly assisted housing for low-income households. In recent years, however, the definition in practice has been broadened to include units with "shallow" subsidies which are intended for moderate-income households.

8 Residential intensification refers to increasing the occupancy of existing homes, or "infilling" vacant lots with new units, or converting non-residential buildings to dwelling units. Down-zoning occurs when public agencies change existing zoning to a category that requires lower densities. Inclusionary zoning compels developers, in return for permission to build, to include specified uses (e.g. low-cost housing) within their projects. Zoning bonuses are used to extract public benefits (e.g. day care facilities, parks, public parking) in exchange for increased floor space in new developments.

BIBLIOGRAPHY

Adams, T.K., Duncan, G.J., and Rodgers, W.L. (1988) *Persistent Urban Poverty: Prevalence, correlates, and trends*, Ann Arbor: University of Michigan Survey Research Center.

Agnelli, S. (1986) *Street Children: A growing urban tragedy*, a report for the Independent Commission on International Humanitarian Issues, London: Weidenfeld and Nicolson.

Ambrosio, E. and Baker, D. (1992) "The Street Health Report: a study of the health status and barriers to health care of homeless women and men in the City of Toronto," Toronto: Street Health.

Anderson, I. (1993) "Housing policy and street homelessness in Britain," *Housing Studies* 8, 1: 17–28.

Apgar, W.C. (1989) *The Nation's Housing: A review of past trends and future prospects for housing in America*, MIT Housing Policy Project HP No. 1, Cambridge, Mass.: MIT Center for Real Estate Development.

Appelbaum, R.P., Dolny, M., Dreier, P., and Gilderbloom, J.I. (1991) "Scapegoating rent control: Masking the causes of homelessness," *Journal of the American Planning Association* 57, 2: 153–164.

Archbishop of Canterbury (1985) *Faith in the City: The report of the Archbishop of Canterbury's Commission on Urban Priority Areas*, London: Church House Publishing.

Armstrong, P. and Armstrong, H. (1994) *Taking Care: Warning signals for Canada's health system*, Toronto: Garamond.

Association of Municipalities of Ontario (1986) *Report on Deinstitutionalization*, Toronto.

Audit Commission (1989) *Housing the Homeless: The local authority role*, London: HMSO.

Bacher, J.C. and Hulchanski, J.D. (1987) "Keeping warm and dry: the policy response to the struggle for shelter among Canada's homeless, 1900–1960," *Urban History Review* 16, 2: 147–163.

Bachrach, L. (ed.) (1983) *Deinstitutionalization*, San Francisco: Jossey-Bass.

Baker, S.G. (1994) "Gender, ethnicity and homelessness: accounting for demographic diversity on the streets," *American Behavioral Scientist* 37, 4: 476–504.

Bane, M.J. and Ellwood, D.T. (1994) *Welfare Realities: From rhetoric to reform*, Cambridge, Mass.: Harvard University Press.

Bane, M.J. and Jargowsky, P.A. (1988) *Urban Poverty Areas: Basic questions concerning prevalence, growth and dynamics*, Cambridge, Mass.: John F. Kennedy School of Government.

Barak, G. (1991) *Gimme Shelter: A social history of homelessness in contemporary America*, New York: Praeger.

Bard, M. (1994) *Organizational Community Responses to Domestic Abuse and Homelessness*, New York: Garland.

Barlow Commission (1940) *Report of the Royal Commission on the Distribution of the Industrial Population*, Cmd. 6153, London: HMSO.

Bassuk, E.L. and Gallagher, E.M. (1990) "The impact of homelessness on children," in Boxill, N.A. (ed.) *Homeless Children: The watchers and the waiters*. New York: The Haworth Press, 19–33.
Bassuk, E.L. and Rosenberg, L. (1988) "Why does family homelessness occur?" *American Journal of Public Health* 78: 783–788.
Bauer, C. (1934) *Modern Housing*, Boston: Houghton Mifflin.
Bean, G.J., Stefl, M.E. and Howe, S.R. (1987) "Mental health and homelessness," *Social Work* September/October: 411–416.
Beardshaw, V. (1988) *Last on the List: Community services for people with physical disabilities*, London: The King's Fund Institute.
Belcher, J.R. and DiBlasio, F.A. (1990) *Helping the Homeless: Where do we go from here?* Lexington, Mass.: Lexington Books.
Belcher, J.R. and Toomey, B. (1988) "Relationships between the deinstitutionalization model, psychiatric disability, and homelessness," *Health and Social Work* (Spring) 13, 2: 145–153.
Beveridge, W. (1942) "The Beveridge Report," *Social Insurance and Allied Services*. London: HMSO.
Betz, H.-G. (1993) "The new politics of resentment: radical right-wing populist parties in Western Europe," *Comparative Politics* 25, 4: 413–427.
Black, D. (1980) *Black Report: Inequalities in health*, London: HMSO; Penguin edition, 1982.
Blackwell, J. (1988) "Is there a need for change in housing policy?" in Blackwell, J. and Kennedy, S. (eds) *Focus on Homelessness*, Dublin: Focuspoint.
Blankertz, L.E. and Cnaan, R.A. (1994) "Assessing the impact of two residential programs for dually diagnosed homeless individuals," *Social Service Review* (December): 537–560.
Blau, J. (1992) *The Visible Poor: Homelessness in the United States*, New York: Oxford University Press.
Boston Foundation (1989) *In the Midst of Plenty: A profile of Boston and its poor*, Boston: The Boston Foundation.
Bosworth, W. (1987) "New partnerships – building for the future," in Lang-Runtz, H. and Ahem, D.C. (eds) *Proceedings of the Canadian Conference to Observe the International Year of Shelter for the Homeless*, Ottawa, September, 84.
Bowley, M. (1945) *Housing and the State, 1919–1944*, London: Allen and Unwin.
Bratt, R., Hartman, C., and Meyerson, A. (eds) (1986) *Critical Perspectives on Housing*, Philadelphia: Temple University Press.
Breakey, W. R., Fisher, P.J., Kramer, M., Nestadt, G., Romanski, A.J., Ross, A., Royall, R.M., and Stine, O.C. (1989) "Health and mental problems of homeless men and women in Baltimore," *Journal of the American Medical Association* 262: 1352.
Breitbart, M. (1990) "Quality Housing for Women and Children," *Canadian Woman Studies* 11, 2: 18–24.
Brenton, M. (1985) *The Voluntary Sector in British Social Services*, London: Longman.
Bresalier, M. and O'Donnell, S. (1995) "Editorial essay," *Undercurrents* 7, Toronto: York University Faculty of Environmental Studies, 2–3.
Brickner, P. W., Scharer, L.K., Conanan, B., Elvy, A., and Savarese, M. (eds) (1985) *Health Care Of Homeless People*, New York: Springer.
Brinkman, H. J. (1990) "Why estimates of the incidence of poverty differ," Working Paper No. 4, Department of International Economic and Social Affairs, United Nations.
Brown, J. (1995) "Shelter for homeless women: a comparison of Toronto and Nairobi," unpublished MES thesis, York University, Toronto.
Brown, L. (1987) *When Freedom Was Lost: The unemployed, the agitator, and the state*. Montreal: Black Rose Books.
Bublick, R. (1991) "Programs without policy: Winnipeg's street youth," *Canadian Housing*, 8, 3.
Buck, N., Gordon, I., and Young, K. (l987) "London: employment problems and

prospects," in Hausner, V.A. (ed.) *Urban Economic Change: Five city studies*, Oxford: Clarendon Press, 99–131.

Burghardt, S. and Fabricant, M. (1987) *Working Under the Safety Net: Policy and practice with the new American poor*, Newbury Park: Sage Publications, A Sage Human Services Guide #47.

Burt, M. (1992) *Over the Edge: The growth of homelessness in the 1980s*, New York: Russell Sage Foundation.

Burt, M. and Cohen, B.E. (1989) *America's Homeless: Numbers, characteristics, and the programs that serve them*, Washington, D.C.: Urban Institute Press.

Canada (1919) *Report of the Royal Commission to Enquire into Industrial Relations in Canada together with a Minority Report*, Ottawa.

—— (1940) *Report of the Royal Commission on Dominion–Provincial Relations*, Ottawa.

Canada Mortgage and Housing Corporation (1989) *Maintaining Seniors' Independence*, Ottawa: CMHC.

—— (1990) *Evaluation of the Public Housing Program*, Ottawa: CMHC.

—— (1994) *Research and Development Highlights*, Ottawa: CMHC.

Canadian Association for Community Living (1989) *Report* February 17.

Canadian Association of Food Banks (1991) *HungerCount 1990*, Toronto.

—— (1993) *HungerCount 1992*, Toronto.

Canadian Bar Association (1995) *Aboriginal Law: Issues that matter in the 1990s*, Toronto: Canadian Bar Association.

Canadian Council on Social Development (1985) *Deinstitutionalization: Costs and effects*, Toronto.

—— (1987) "Homelessness in Canada: the report of the national inquiry," *Social Development Overview* 5, 1, Ottawa: CCSD.

Canadian Human Rights Commission (1989) *Annual Report 1989*, Ottawa: Canadian Human Rights Commission.

Carter, T. and McAfee, A. (1990) "The municipal role in housing the homeless and poor," in Fallis, G. and Murray, A. (eds) *Housing the Homeless and Poor: New partnerships among the private, public and third sectors*, Toronto: University of Toronto Press, 227–262.

Castells, M. (1991) *The Informational City*. Oxford: Basil Blackwell.

Center on Budget and Policy Priorities (1991) *The States and the Poor: How budget decisions in 1991 affected low-income people*, Washington, D.C.

Central Statistical Office (1990) *Annual Abstract of Statistics*, London: HMSO.

—— (1992a) *Annual Abstract of Statistics*, London: HMSO.

—— (1992b) *Social Trends 22*, London: HMSO.

—— (1995) *Social Trends 25*, London: HMSO.

City of Boston (1986) *Rooms for Rent: A study of lodging housing in Boston*, Boston.

—— (1995) *State of Homelessness in the City of Boston, Winter 1994–95*, Boston: Emergency Shelter Commission.

City of Toronto, Department of Planning and Development (1986) "Housing in the Toronto region," *City Planning* Fall: 42–46.

—— (1987) *Report of the Inquiry into the Effects of Homelessness on Health*, Toronto: Department of the City Clerk.

—— (1990) *Homeless Not Helpless*. Toronto: Healthy City Office.

Coalition for the Homeless (1983) *Cruel Brinkmanship: Planning for the homeless*, New York: Coalition for the Homeless.

Cochrane, A. (1985) "The attack on local government: what it is and what it isn't," *Critical Social Policy* 4, Spring: 44–61.

Commission of the European Community (1990) *Final Report of the Second European Poverty Programme, 1985–89*, Brussels: Commission of the European Community.

Commission on Private Philanthropy and Public Needs (1975) *Giving in America: Toward a stronger voluntary sector*, Washington, D.C.

Commission for Racial Equality (1984) *Hackney Housing Investigated: Summary of a formal investigation report*. London: CRE.

—— (1990) *"Sorry It's Gone": Testing for racial discrimination in the private rented housing sector*, London: CRE.

—— (1993) *Annual Report*, London: HMSO.

Conrad, K.J., Hultmand, C.I., and Lyons, J.S. (eds) (1993) *Treatment of the Chemically Dependent Homeless: Theory and implementation in fourteen American projects*, New York: Haworth Press.

Conway, J. (1988) *A Prescription for Ill Health: The crisis for homeless families*, London: Shelter.

Conway, J. and Kemp, P. (1985) *Bed and Breakfast: Slum housing of the Eighties*, London: Shelter.

Coulson, C. (1987) "The $37,000 slum: why New York City pays $1,800 a month to house one family in one miserable room," *New Republic*, January 19, 15–16.

Crane, H. (1990) *Speaking from Experience: Working with homeless families*, London: Bayswater Hotel Homelessness Project.

Cross, M. (ed.) (1992) *Ethnic Minorities and Industrial Change in Europe and North America*, Cambridge: Cambridge University Press.

Cross, M. and Waldinger, R. (1992) "Migrants, minorities, and the ethnic division of labor," in Fainstein, S., Gordon, I., and Harloe, M. (eds) *Divided Cities: New York and London in the contemporary world*. Oxford: Blackwell, 151–171.

Culhane, D., Dejowski, E., Ibanez, J., Needham, E., and Macchia, I. (1993) "Public shelter admission rates in Philadelphia and New York City: the implications of turnover for sheltered population counts," Washington, D.C.: Office of Housing Research, Fannie Mae working paper.

Cullingworth, J.B. (1993) *The Political Culture of Planning: American land use planning in comparative perspective*, New York and London: Routledge.

Daly, G. (1988) *A Comparative Assessment of Programs Dealing with the Homeless Population in the United States, Canada, and Britain*, Ottawa: Canada Mortgage and Housing Corporation.

—— (1989) "Homelessness and health: views and responses from the United States, Canada and the UK," *Health Promotion Journal* 4, 2: 115–128.

—— (1990) "Programs dealing with homelessness in the United States, Canada, and Britain," in Momeni, J.A. (ed.) *Homelessness in the United States Vol. II: Data and Issues*, New York: Greenwood Press, 133–152.

—— (1993a) "Place making by residents: the case of self-help housing," *Open House International*, 18, 1: 20–27.

—— (1993b) "The state response to homelessness," in Frisken, F. (ed.) *The Changing Canadian Metropolis*, University of California at Berkeley Press, and the Canadian Urban Institute.

—— (1996) "Migrants and gate-keepers: the links between homelessness and immigration in Western Europe," *Cities* 13, 1: 11–23.

Daly, G. and Muirhead, B. (1993) *Report on Poverty and Social Policy in Industrial Countries*, Ottawa: International Development Research Centre.

Dangschat, J. and Ossenbrugge, J. (1990) "Hamburg: crisis management, urban regeneration, and social democrats," in Judd, D. and Parkinson, M. (eds) *Leadership and Urban Regeneration: Cities in North America and Europe*, Urban Affairs Annual Reviews Vol. 37, London: Sage, 86–105.

Dear, M.J. and Wolch, J.R. (1987) *Landscapes of Despair: From deinstitutionalization to homelessness*, Princeton: Princeton University Press.

Department of Environment (1977) *Code of Guidance to England and Wales Housing (Homeless Persons) Act 1977*, London: HMSO.

—— (1991) *HIP Returns from Local Authorities*, London: HMSO.

—— (1992) *Housing and Construction Statistics*, London: HMSO.

—— (1994a) *Homelessness Statistics*, London: HMSO.

—— (1994b) "Access to Local Authority and Housing Association tenancies: a consultation paper," London: Department of Environment.

—— (1995) *Our Future Homes: The government's housing policies for England and Wales*, Cm 2901, London: HMSO.

Department of Health, Child Care Division (1990) *Survey of Children in Care of Health Boards in 1988* Vol. I, Dublin.
Department of Health and Social Security (1985) *Green Paper: Reform of Social Security*, Cmnd. 9517, London: HMSO.
—— (1987) "Duration of spell of unemployment," in *Unemployment Benefit Summary Sheets*, London: HMSO.
Dhillon-Kashyap, P. (1994) "Black women and housing", in Gilroy, R. and Woods, R. (eds) *Housing Women*, London: Routledge, 101–126.
DiBlasio, F.A. and Belcher, J.R. (1995) "Gender differences among homeless persons: special services for women," *American Journal of Orthopsychiatry*, 65, 1: 131–137.
Dillon, B. and O'Brien, L. (1982) *Private Rented, the Forgotten Sector*, Dublin: Threshold.
Disraeli, B. (1845/1980) *Sybil: or, the Two Nations*, New York: Penguin.
Dixon, L., Krauss, N., and Lehman, A. (1994) "Consumers as service providers: the promise and challenge," *Community Mental Health Journal* 30, 6: 615–634.
Dolbeare, C. and Kaufman, T. (1995) *Out of Reach: Why everyday people can't find affordable housing*, Washington, D.C.: Low Income Housing Information Service.
Donnison, D. and Ungerson, C. (1982) *Housing Policy*, Harmondsworth: Penguin.
Donovan, K. (1995) "Housing millions down drain," *Toronto Star*, May 20, 1995: A1.
Dovey, K. (1985) "Home and homelessness," in Altman, I. and Werner, C. (eds) *Home Environments*, New York: Plenum Press.
Drake, R.E., Becker, D.R., Biesanz, J.C., Torrey, W.C., McHugo, G.J., and Wyzik, P.F. (1994) "Rehabilitative day treatment: I. Vocational outcomes," *Community Mental Health Journal* 30, 5: 519–530.
Economic Council of Canada (1991) *Canadian Unemployment: Lessons from the 80s and challenges for the 90s*, Ottawa: Economic Council of Canada.
Edelman, M.R. (1987) *Families in Peril: An agenda for social change*, Cambridge, Mass.: Harvard University Press.
Ellwood, D.T. (1988) *Poor Support: Poverty in the American family*, New York: Basic Books.
—— (1993) "The changing structure of American families," *Journal of the American Planning Association* 59, 1: 3–8.
—— (1994) "Reducing poverty by replacing welfare," in Bane, M.J. and Ellwood, D.T., *Welfare Realities: From rhetoric to reform*, Cambridge, Mass.: Harvard University Press, 143–162.
Engeland, J. (1990–91) "Canadian renters in core housing need," *Canadian Housing*, 7, 4: 6–10.
Enoch, Y. (1992) "The intolerance of a tolerant people: ethnic relations in Denmark," *Ethnic and Racial Studies* 17: 282–300.
Equal Opportunities Commission (1993) *Women and Men in Britain, 1993*, Manchester: EOC.
Evans, A. and Duncan, S. (1988) *Responding to Homelessness: Local authority policy and practice*, London: HMSO.
Fainstein, S., Gordon, I., and Harloe, M. (eds) (1992) *Divided Cities: New York and London in the contemporary world*, Oxford: Blackwell.
Federation of Canadian Municipalities (1991) "Housing and homelessness," Vancouver.
Ferrand-Bechmann, D. (1988) "Homeless in France: public and private policies," in Friedrichs, J. (ed.) *Affordable Housing and the Homeless*, Berlin: Walter de Gruyter.
First, R.J., Roth, D., and Arewa, B.D. (1988) "Homelessness: understanding the dimensions of the problem for minorities," *Social Work*, 33, 2: 120–124.
Fischer, P.J. (1988) "Criminal activity among the homeless: a study of arrests in Baltimore," *Hospital and Community Psychiatry* 39: 46–51.
Fischer, P.J. and Breakey, W.R. (1991) "The epidemiology of alcohol, drug, and mental disorders among homeless persons," *American Psychologist*, 46, 11: 1115–1128.
Fisher, K. and Collins, J. (eds) (1993) *Homelessness, Health and Welfare Provision*, London and New York: Routledge.

Foodshare Metro Toronto (1992) *Twice Vulnerable: A preliminary profile on access to emergency food assistance for Metro's multicultural communities*, Toronto.
Foot, M. (1973) *Aneurin Bevan*, London: Davis-Poynton.
Forrest, R. and Murie, A. (1990) *Residualisation and Council Housing: A statistical update*, Working Paper 91, School for Advanced Urban Studies, University of Bristol.
Förster, M. F. (1994) "Measurement of low incomes and poverty in a perspective of international comparisons," *Occasional Paper No. 14*, Paris: OECD.
Fox, E.R. and Roth, L. (1989) "Homeless children: Philadelphia as a case study," *The Annals of American Academy of Political and Social Science* 506, November: 141–151.
Fraser, R. (1986) *Filling the Empties: Short life housing and how to do it*, London: Shelter.
Freeman, G.P. (1992) "Migration policy and politics in the receiving states," *International Migration Review* 26, 4: 1144–1167.
Freeman, R.B. and Hall, B. (1986) *Permanent Homelessness in America*, Cambridge, Mass.: National Bureau of Economic Research.
Frieden, B.J. and Kaplan, M. (1977) *The Politics of Neglect: Urban aid from model cities to revenue sharing*, 2nd edn, Cambridge, Mass.: MIT Press.
Friedrichs, J. (1988) "Large new housing estates: the crisis of affordable housing," in Friedrichs, J. (ed.) *Affordable Housing and the Homeless*, Berlin: Walter de Gruyter, 89–102.
Galbraith, J.K. (1992) "The tyranny of the contented," *The Washington Monthly* (June): 29–31.
Garrow, I. (1986) "Revisiting homelessness on Skid Row," unpublished M.E.S. thesis, York University, Toronto.
Gelberg, L., Linn, L.S., and Leake, B.D. (1988) "Mental health, alcohol and drug use, and criminal history among homeless adults," *American Journal of Psychiatry* 145: 191–196.
Gerstein, R. (1984) *Final Report of the Mayor's Action Task Force on Discharged Psychiatric Patients*, Toronto: Office of the Mayor.
Giamo, B. and Grunberg, J. (1992) *Beyond Homelessness: Frames of reference*, Iowa City: University of Iowa Press.
Gilbert, B. B. (1970) *British Social Policy, 1914–1939*, London: Batsford.
Gilroy, R. (1994) "Women and owner occupation in Britain: first the prince, then the palace?" in Gilroy, R. and Woods, R. (eds) *Housing Women*, London: Routledge.
Glastonbury, B. (1971) *Homeless Near a Thousand Homes: A study of families without homes in South Wales and the West of England*, London: George Allen and Unwin.
Golden, S. (1992) *The Women Outside: Myths and meanings of homelessness*, Berkeley: University of California Press.
Goodwin, D.W. (1994) *Alcoholism: The facts*, Oxford: Oxford University Press.
Gordon, L. (ed.) (1990) *Women, the State, and Welfare*, Madison: University of Wisconsin Press.
Government of Ontario (1990) *Report of the Standing Committee on Social Development*, Toronto.
Green, J. (1987) "The politics of homelessness," in Lang-Runtz, H. and Ahem, D.C. (eds) *Proceedings of the Canadian Conference to Observe the International Year of Shelter for the Homeless*, Ottawa, 63–66.
Greve, J. (1971) *Homelessness in London*, Edinburgh: Scottish Academic Press.
Greve, J. with Currie, E. (1990) *Homelessness in Britain*, York: Joseph Rowntree Trust.
Gueron, J.M. (1991) *From Welfare to Work*, New York: Russell Sage Foundation.
Guest, D. (1985) *The Emergence of Social Security*, 2nd edn, Vancouver: University of British Columbia Press.
Haar, C. (ed.) (1984) *Cities, Law and Social Policy: Learning from the British*, Lexington, Mass.: Lexington Books.
Hackney, R. (1988) "Community architecture," *Open House International* 13, 3: 41–48.
Hambleton, R. (1990) "Future directions for urban government in Britain and America," *Journal of Urban Affairs* 12, 1: 75–94.

Hamburger, C. (1992) "The development of policy on denizens in Denmark," *New Community* 18, 2: 293–310.
Hargreaves, A.G. (ed.) (1987) *Immigration in Post-war France: A documentary anthology*, London: Methuen.
Harloe, M., Marcuse, P., and Smith, N. (1992) "Housing for people, housing for profits," in Fainstein, S., Gordon, I., and Harloe, M. (eds) *Divided Cities: New York and London in the contemporary world*. Oxford: Blackwell, 175–202.
Harris, M. (1991) *Sisters of the Shadow*, Norman: University of Oklahoma Press.
Haughland, G., Craig T.J., Goodman, A.B., and Siegel, C. (1983) "Mortality in the era of deinstitutionalization," *American Journal of Psychiatry* 23: 377–385.
Havel, V. (1992) *Summer Meditations*, Toronto: Knopf.
Hayes, R. (1989) *Current Biography*, April, New York: H.W. Wilson.
Heisenberg, W. (1970) *The Physicist's Conception of Nature* (translated from the German by A.J. Pomerans) Westport, Conn.: Greenwood Press.
Hill, O. (1884) *Homes of the London Poor*. London: Macmillan.
Hjarno, J. (1992) "Migrants and refugees on the Danish labour market," *New Community* 18, 1: 75–87.
Holmans, A. (1991) "The 1977 National Housing Policy Review in retrospect," *Housing Studies* 6, 3.
Hopper, K. and Hamberg, J. (1986) "The making of America's homeless: from skid row to new poor, 1945–1984," in Bratt, R., Hartman, C., and Meyerson, A. (eds) *Critical Perspectives on Housing*, Philadelphia: Temple University Press.
Housing Support Team (1987) *Briefing Papers*, London: Housing Support Team.
Howe, F. (1913) *European Cities at Work*, New York: C. Scribner's Sons.
Howe, N. and Longman, P. (1992) "The next New Deal," *The Atlantic* 269, 4: 88–99.
Hulchanski, J.D., Eberle, M., Olds, K., and Stewart, D. (1991) *Solutions to Homelessness: Vancouver case studies*, a report prepared by the Centre for Human Settlements, University of British Columbia for the Canada Mortgage and Housing Corporation.
Independent Commission on International Humanitarian Issues (1986) *Street Children: A growing urban tragedy*, London: Weidenfeld and Nicolson.
Institute of Medicine (1988) *Homelessness, Health, and Human Needs*, Washington, D.C.: National Academy Press.
Irwin, C. (1988) "Lords of the Arctic: wards of the state. The growing Inuit population, Arctic resettlement and their effects on social and economic change," Ottawa: Health and Welfare Canada.
Jaffe, D.J. and Howe, E. (1988) "Agency-assisted shared housing: the nature of programs and matches," *The Gerontologist* 28, 3: 318–324.
Jencks, C. (1994) *The Homeless*, Cambridge, Mass.: Harvard University Press.
Johnson, N. (1987) *The Welfare State in Transition*, London: Wheatsheaf.
Kanter, A.S. (1989) "Homeless but not helpless: legal issues in the care of homeless people with mental illness," *Journal of Social Issues* 45, 3: 91–104.
Katz, M. (1989) *The Undeserving Poor: From the war on poverty to the war on welfare*, New York: Pantheon.
Kaufman, N. (1987) "The Massachusetts approach to homelessness," unpublished paper presented to Ottawa conference for the International Year of Shelter for the Homeless, October.
Kazim, P. (1991) "Racism is no paradise!" *Race and Class* 32, 3: 84–89.
Kelleher, P. (1990) *Housing and settling homeless people in Dublin City: The implementation of the Housing Act 1988*, Dublin: Focus Point.
Keller, C. (1986) *From a Broken Web: Separation, sexism, and self*, Boston: Beacon Press.
Kelly, J.T. (1985) "Trauma: with the example of San Francisco's shelter programs," in Brickner, P.W., Scharer, L.K., Conanan, B., Elvy, A., and Savarese, M. (eds) *Health Care of Homeless People*, New York: Springer, 77–91.
Kemeny, J. (1992) *Housing and Social Theory*, London and New York: Routledge.
Kennedy, W. (1983) *Ironweed*, New York: Viking Press.
Kerr, G. (1993) "What's going on with Canadian food banks?" *Perception*, 17, 2: 17–18.

Keyes, L. (1988) "Housing and the homeless," HP#15, MIT Housing Policy Project Working Papers, MIT Center for Real Estate Development.

King, A. (1995) "Re-presenting world cities: cultural theory/social practice," in Knox, P.L. and Taylor, P.J. (eds) *World Cities in a World-System*, Cambridge University Press, 215–231.

Kleinman, M. (1992) "Policy responses to changing housing markets," in Hartman, C. and Rosenberg, S. (eds) *Housing Issues of the 1990s*, New York: Praeger, 43–65.

Koegel, P. and Burnam, M.A. (1988) *Alcoholism among Homeless Adults in the Inner city of Los Angeles*, Los Angeles: Planning Institute, University of Southern California.

Kolodny, R. (1986) "The emergence of self-help as a housing strategy for the urban poor," in Prak, N. and Priemus, H. (eds) *Post-war Public Housing in Trouble*, Delft: Delft University Press, 11–18.

Kozol, J. (1988) *Rachel and her Children: Homeless families in America*, New York: Crown.

Krauthammer, C. (1993) "Defining deviancy up," *New Republic*, No. 4114, November 22: 20–25.

Lang-Runtz, H. and Ahem, D.C. (1987) *Proceedings of the Canadian Conference to Observe the International Year of Shelter for the Homeless*, Ottawa.

Laven, G.T. and Brown, K.C. (1985) "Nutritional status of men attending a soup kitchen: a pilot study," *American Journal of Public Health* 75, 8: 875–878.

Laws, G. (1992) "Emergency shelters in urban areas: serving the homeless in metropolitan Toronto," *Urban Geography* 13, 2: 99–126.

Layton-Henry, Z. (1990) *The Political Rights of Migrant Workers in Western Europe*, London: Sage.

Leavitt, J. and Saegert, S. (1989) *From Abandonment to Hope: Community households in Harlem*, New York: Columbia University Press.

Leonard, P. and Lazere, E. (1992) *A Place to Call Home: The low-income housing crisis in 44 major metropolitan areas*, Washington, D.C.: Center on Budget and Policy Priorities.

Levine, I.S. and Stockdill, J.W. (1986) "The mentally ill and the homeless: a national problem," in Jones, B. (ed.) *Treating the homeless: Urban psychiatry's challenge*, Washington, D.C.: American Psychiatric Association Press.

Levitan, S.A. and Gallo, F. (1988) *A Second Chance: Training for jobs*, Kalamazoo, Mich.: W.E. Upjohn Institute for Employment Research.

Lewis, J. (1993) "Developing the mixed economy of care: emerging issues for voluntary organizations," *Journal of Social Policy* 22, 2: 173–192.

Ley, D. (1985) *Gentrification in Canadian Cities: Patterns, analysis, impact and policy*, Vancouver: University of British Columbia, Department of Geography.

—— (1993) *The Changing Geography of Canadian Cities*, Montreal: McGill–Queen's University Press.

Liebow, E. (1993) *Tell Them Who I Am: The lives of homeless women*, New York: Free Press.

Lifton, R.J. (1992) "Victims and survivors," in Giamo, B. and Grunberg, J. *Beyond Homelessness: Frames of reference*, Iowa City: University of Iowa Press, 129–156.

Link, B., Susser, E., Stueve, A., Phelan, J., Moore, R., and Struening, E. (1994) "Lifetime and five-year prevalence of homelessness in the United States," *American Journal of Public Health* 84, 12: 1907–1912.

Lipset, S. (1990) *Continental Divide: The values and institutions of the United States and Canada*, New York: Routledge.

Livingston, J. (1994) *Rogue Primate*, Toronto: Key Porter.

Lloyd, C. and Waters, H. (1991) "France: one culture, one people?" *Race and Class* 32, 3: 49–65.

Logan, J., Taylor-Gooby, P., and Reuter, M. (1992) "Poverty and income inequality," in Fainstein, S., Gordon, I., and Harloe, M. (eds) *Divided Cities: New York and London in the contemporary world*. Oxford: Blackwell, 129–150.

London Housing Unit (1993) *Housing the Poorer Sex*, London: London Housing Unit.

Low Income Housing Information Service (1994) *Setting Goals: HUD's program and management plan for 1994*, Washington, D.C.: Low Income Housing Information Service.

Lowell, J.S. (1884) *Public Relief and Private Charity*, New York: Putnam's.

Mabardi, J.-F. (1985) "Belgian post-war housing in trouble," in Prak, N. and Priemus, H. (eds) *Post-war Public Housing in Trouble*, Delft: Delft University Press, 73–84.

Maclennan, D. and Gibb, K. (1990) "Housing finance and subsidies in Britain after a decade of 'Thatcherism'," *Urban Studies* 27, 6: 905–918.

Maclennan, D., Gibb, K., and More, A. (1990) *Fairer Subsidies, Faster Growth: Housing, government and the economy*, York: Joseph Rowntree Foundation.

Mangen, S. (1992) "Marginalisation and inner city Europe," in Hantrais, L., Mangen, S., and O'Brien, M. (eds) *Dualistic Europe: Marginalisation in the EC of the 1990s*, Loughborough: The Cross-National Research Group.

Marcuse, P. (1987) "Why are they homeless?" *The Nation*, April 4: 426–429.

Marin, P. (1987) "Helping and hating the homeless: the struggle at the margins of America," *Harper's* (January): 39–49.

Martin, M.A. (1986) *The Implications of NIMH-supported research for homeless mentally ill racial and ethnic minority persons*, New York: Hunter College of Social Work.

Mason, D. and Jewson, N. (1992) "'Race', equal opportunities policies and employment practice: reflections on the 1980s, prospects for the 1990s," *New Community* 19, 1: 99–112.

McCay, B.J. and Acheson, J. M. (eds) (1987) *The Question of the Commons: The culture and ecology of communal resources*. Tucson: University of Arizona Press.

McClain, J. with Doyle C. (1983) *Women and Housing: Changing needs and the failure of policy*, Ottawa: Canadian Council on Social Development.

McGregor, J. (1987) "New partnerships – building for the future," in Lang-Runtz, H. and Ahem, D.C. (eds) *Proceedings of the Canadian Conference to Observe the International Year of Shelter for the Homeless*, Ottawa, September, 96.

McGurk, P. and Raynsford, N. (1984) *A Guide to Housing Benefits*, London: SHAC.

McLeod, B. (1993) *Toronto Star*, October 11.

McMaster, M. and Browne, N. (1973) *A Study on Roomers*, Toronto.

McQuaig, L. (1992) "The fraying of our social safety net," *Toronto Star* Insight Series, Nov. 8.

Merckx, F. and Fekete, L. (1991) "Belgium: the racist cocktail," *Race and Class* 32, 3: 67–78.

Merelman, R.M. (1991) *Partial Visions: Culture and politics in Britain, Canada, and the United States*, Madison: University of Wisconsin Press.

Mesler, B. (1995) "The homeless learn how to hit back," *Third Force* (May/June): 1–2.

Mezzina, R., Mazzuia, P., Vidoni, D., and Impagnatiello, M. (1992) "Networking consumers' participation in a community mental health service: mutual support groups, 'citizenship' and coping strategies," *The International Journal of Social Psychiatry* 38, 1: 68–73.

Milanesi, G. (1992) "Scale and definition of homelessness in Europe," in Yanetta, A. and Aldridge, R. (eds) *Housing, Homelessness and Europe*, a report based on papers given at a conference in Edinburgh, Edinburgh: Institute of Housing in Scotland and Scottish Council for Single Homelessness, 13–26.

Miller, M. (1990) *Bed & Breakfast: Women and homelessness today*, London: The Women's Press.

MIND (1987) *Homelessness and the Plight of Mentally Ill People*, London: MIND.

Mingione, E. and Morlicchio, E. (1993) "New forms of urban poverty in Italy: risk path models in the North and South," *International Journal of Urban and Regional Research* 17, 3: 413–427.

Momeni, J. A. (ed.) (1989) *Homelessness in the United States. Vol. I: State surveys*, New York: Greenwood Press.

Momeni, J. A. (ed.) (1990) *Homelessness in the United States. Vol. II: Data and issues*, New York: Greenwood Press.

Morris, L. (1994) *Dangerous Classes: The underclass and social citizenship*, London and New York: Routledge.
Mulburn, N.G. and Booth, J.A. (1989) "Socio-dynamics, homeless state, and mental health characteristics of women in shelters: preliminary findings," *Urban Research Review*, Washington, D.C.: Howard University Institute for Urban Affairs, 12, 2.
Municipality of Metropolitan Toronto (1987) "Emergency shelters for the homeless," Community Services Department, Hostel Operations, September 25.
—— (1993) "Trends in utilization of hostels," Toronto.
—— (1995) "Trends in utilization of hostels," Toronto.
Murray, C. (1984) *Losing Ground: American social policy, 1950–1980*, New York: Basic Books.
Muus, P. (1991) *Netherlands 1991: A report for the OECD*, Amsterdam: Institute for Social Geography.
Myrdal, G. (1944) *An American Dilemma: The Negro problem and modern democracy*, New York: Harper.
National Advisory Commission on Civil Disorders (1968) *Report of the National Advisory Commission on Civil Disorders*, New York: Bantam Books.
National Commission on Urban Problems (1969) *Building the American City*, New York: Praeger.
National Council of Welfare (1992) *Poverty Profile 1980–1990*, Ministry of Supply and Services Canada, Ottawa.
National Research Council (1993) *The Social Impact of AIDS in the United States*, Washington D.C.: National Academy Press.
Navarro, V. (1985) "The public/private mix in the funding and delivery of health services: an international survey," *American Journal of Public Health* 75, 11: 1318–1320.
Nicholas, H.G. (1986) *The Nature of American Politics*, 2nd edn, Oxford: Oxford University Press.
Niner, P. (1989) *Homelessness in Nine Local Authorities: Case studies of policy and practice*, London: HMSO.
Niner, P. and Maclennan D. (1990) *Inquiry into British Housing*, York: Joseph Rowntree Foundation.
North, C. and Smith, E. (1993) "A comparison of homeless men and women: different populations, different needs," *Community Mental Health Journal* 29, 5: 423–430.
North London Education Project (1987) *Report, 1984–86*, London: North London Education Project.
Oberle, P. R. (1993) *The Incidence of Family Poverty on Canadian Indian Reserves: Quantitative analysis and socio-demographic research*, Management Information and Analysis Branch, Department of Indian Affairs and Northern Development.
Odekirk, J. (1992) "Food banks," *Canadian Social Trends*, Statistics Canada.
Office of Population Censuses and Surveys (OPCS) (1991) *Preliminary Report for England and Wales*, London: HMSO.
Office of Population Censuses and Surveys (1993) *1991 Census: Report for Great Britain*, London: HMSO.
Ontario Ministry of Community and Social Services (1988) *Report of the Standing Committee on Social Development*, Toronto.
Ontario Ministry of Housing (1989) "Housing policy statement," Toronto.
Orloff, A.S. (1993) *The Politics of Pensions: A comparative analysis of Britain, Canada, and the United States, 1880–1940*, Madison: University of Wisconsin Press.
Orwell, G. (1937) *The Road to Wigan Pier*, London: Gollancz.
Ossenbrugge, J. (1990) "Hamburg: crisis management, urban regeneration, and social democrats," in Judd, D. and Parkinson, M. (eds) *Leadership and Urban Regeneration: Cities in North America and Europe*, Urban Affairs Annual Reviews 37, London: Sage 86–105.
Peach, C. and Byron, M. (1993) "Caribbean tenants in council housing," *New Community* 19, 3: 410–423.

Plamondon, J. (1985) "Deinstitutionalization: its implications and requirements," in *Deinstitutionalization: Costs and effects*, proceedings of an invitational symposium, Ottawa: Canadian Council on Social Development, 19–26.

Poinsot, M. (1993) "Competition for political legitimacy at local and national levels among young North Africans in France," *New Community* 20, 1: 79–92.

Pomeroy, S. (1989) "The recent evolution of social housing in Canada," *Canadian Housing* 6, 9: 6–13.

Poor Law Commissioners (1834) *Report of H.M. Commission for Inquiring into the Administration and Practical Operation of the Poor Laws*, London, Parliamentary Papers, XXVII.

Prigogine, I. and Stengers, I. (1984) *Order out of Chaos: Man's new dialogue with nature*, Toronto: Bantam Books.

Public Record Office (1920) "Revolutionaries and the need for legislation," CAB 24/97, CB 544, London: Directorate of Intelligence.

Rapoport, A. (1985) "Thinking about home environments: a conceptual framework," in Altman, I. and Werner, C.M. (eds) *Home Environments*, New York: Plenum Press, 255–286.

Räthzel, N. (1990). "Germany: one race, one nation?" *Race and Class* 32, 3: 31–48.

Raubeson, A. (1987) "New partnerships – building for the future," in Lang-Runtz, H. and Ahem, D.C. (eds) *Proceedings of the Canadian Conference to Observe the International Year of Shelter for the Homeless*, Ottawa, September, 101–103.

Raynsford, N. (1986) "The 1977 Housing (Homeless Persons) Act," in Deakin, N. (ed.) *Policy Change in Government*, London: RIPA, 35–51.

Rekart, J. (1994) *Public Funds, Private Provision: The role of the voluntary sector*, Vancouver: UBC Press.

Reyes, L.M. and Waxman, L.D. (1989) *A Status Report on Hunger and Homelessness in America's Cities: 1989*, Washington, D.C.: U.S. Conference of Mayors.

Rigby (1985) "A community-based approach to salvaging troubled public housing: tenant management," in Prak, N.L. and Priemus, H. (eds) *Post-war Public Housing in Trouble*, Delft: Delft University Press, 19–34.

Riis, J. (1890/1971) *How the Other Half Lives*, New York: Dover.

—— (1902) *The Battle with the Slum*, New York: Macmillan.

Ringenbach, P.T. (1973) *Tramps and Reformers, 1873–1916: The discovery of unemployment in New York*, Westport, Conn.: Greenwood Press.

Riordan, T. (1987) "Housekeeping at HUD: why the homeless problem could get much, much worse," *Common Cause* (March/April): 26–31.

Roelandt, T. and Veenman, J. (1993) "An emerging ethnic underclass in the Netherlands? Some empirical evidence," *New Community* 19, 1: 129–141.

Ropers, R.M. (1991) *Persistent Poverty: The American dream turned nightmare*, New York: Insight Books, Plenum Press.

Rose, L. (1989) "People with AIDS: international housing initiatives," *Canadian Housing* 6, 4: 64.

Rosenbaum, E. (1987) "New partnerships – building for the future," in Lang-Runtz, H. and Ahem, D.C. (eds) *Proceedings of the Canadian Conference to Observe the International Year of Shelter for the Homeless*, Ottawa, September, 102–103.

Rosenheck, R. and Fontana, A. (1994) "A causal model of homelessness among male veterans of the Vietnam generation," *American Journal of Psychiatry* 151: 421–427.

Ross, D.P. (1990) "The facts on income security, 1990," *Perception*, Ottawa: Canadian Council on Social Development, 8–12.

Ross, D.P. and Shillington, E.R. (1989) *The Canadian Fact Book on Poverty – 1989*, Ottawa: Canadian Council on Social Development.

Ross, D.P., Shillington, E.R., and Lochhead, C. (1994) *The Canadian Fact Book on Poverty – 1994*, Ottawa: Canadian Council on Social Development.

Rossi, P.H. (1989) *Down and Out in America: The origins of homelessness*, Chicago: University of Chicago Press.

Rossi, P.H., Wright, J.D., Fisher, G.A., and Willis, G. (1987) "The urban homeless: estimating composition and size," *Science* 235: 1336–1341.

Rowe, S. (1988) *Social Networks in Time and Space: The case of homeless women in skid row, Los Angeles*, Los Angeles: Planning Institute, University of Southern California.

Rybczynski, W. (1986) *Home: The short history of an idea*. New York: Viking.

Sahlin, I. (1994) "Discipline and border control: strategies for tenant control and housing exclusion," paper presented to the 13th World Congress of Sociology, Bielefeld, Germany, July.

Sarlo, C. A. (1992) *Poverty in Canada*, Vancouver: The Fraser Institute.

Sassen, S. (1991) *The Global City: New York, London, Tokyo*, Princeton, N.J.: Princeton University Press.

Saunders, B. (1986) *Homeless Young People in Britain: The contribution of the voluntary sector*, London: Bedford Square Press.

Scarman, L. (1981) *The Brixton Disorders 10–12 April: Report of an inquiry*, Cmnd. 8427, London: HMSO.

Schain, M.A. (1993) "Policy-making and defining ethnic minorities: the case of immigration in France," *New Community* 20, 1: 59–77.

Schnabel, P. (1992) "Down and out: social marginality and homelessness," *International Journal of Social Psychiatry* 38, 1: 59–67.

Schutt, R.K. (1985) "Boston's homeless: their background, problems and needs," prepared for the Long Island Shelter for the Homeless, Boston.

—— (1992) *Responding to the Homeless: Policy and practice*, New York: Plenum Press.

Sciortine, G. (1991) "Immigration into Europe and public policy: do stops really work?" *New Community* 18, 1: 89–99.

Shalivitz, J., Goulart, M., Dunnigan, K., and Flannery, D. (1990) "Prevalence of STD and HIV in a homeless youth medical clinic in San Francisco," in *Proceedings of the International Conference on AIDS* 6, 3: 231.

Shelter (1994) "Access to local authority and housing association tenancies, a consultation paper: Shelter's response," London: Shelter.

—— (1995) "1995 Housing White Paper: Shelter briefing," London: Shelter.

Shepard, P. (1982) *Nature and Madness*. San Francisco: Sierra Club Books.

Sherman, D. (1992) "The neglected health care needs of street youth," *Public Health Report* 107, 4: 433–440.

Shragge, E. (1990) "Community-based practice: political alternatives or new state forms," in Davies, L. and Shragge, E. (eds) *Bureaucracy and Community*, Montreal: Black Rose Books, 134–178.

Single Displaced Persons' Project (1983) "The case for long-term supportive housing," Toronto.

—— (1987) "From homelessness to home: a case for facilitative management," Toronto.

Slessarev, H. (1988) "Racial tensions and institutional support: social programs during a period of retrenchment," in Weir, M., Orloff, A.S., and Skopcol, T. (eds) *The Politics of Social Policy in the United States*, Princeton, N.J.: Princeton University Press, 357–379.

Smeeding, T.M., O'Higgins, M., and Rainwater, L. (1990) *Poverty, Inequality and Income Distribution in Comparative Perception*, Hemel Hempstead: Harvester Wheatsheaf.

Snow, D.A. and Anderson, L. (1993) *Down on their Luck: A study of homeless street people*, Berkeley: University of California Press.

Snowman, D. (1977) *Britain and America: An interpretation of their culture, 1945–1975*, New York: New York University Press.

Spaull, Sue (1992) "Projects for single homeless people in European cities," in Yanetta, A. and Aldridge, R. (eds) *Housing, Homelessness and Europe*, a report based on papers given at a conference in Edinburgh, Edinburgh: Institute of Housing in Scotland and Scottish Council for Single Homelessness, 39–54.

Specht-Kittler, T. (1992) "Homelessness and the housing crisis in Germany," in Yanetta, A. and Aldridge, R. (eds) *Housing, Homelessness and Europe*, a report based

on papers given at a conference in Edinburgh, Edinburgh: Institute of Housing in Scotland and Scottish Council for Single Homelessness, 27–37.
Sprague, J.F. (1986) *Transitional Housing*, Boston: Women's Institute for Housing and Economic Development.
—— (1991) *More than Housing: Lifeboats for women and children*, London: Butterworth-Heinemann.
Statistics Canada (1990) *Canadian Social Trends* Autumn, No. 17, Ottawa: Statistics Canada.
—— (1992) "The labour market: year-end review," *Canadian Economic Observer*, Ottawa: Statistics Canada.
—— (1995) *Labour Force Annual Averages, 1989–1994*, Ottawa: Statistics Canada.
Streetwise National Coalition (1988) *A National Survey of Young People out of Home in Ireland*, Dublin: Streetwise.
Stricof, R.L., Novick, L.F., and Kennedy, J.T. (1990) "HIV-1 seroprevalence in facilities for runaway and homeless youth adolescents in four states," in *Proceedings of the International Conference on AIDS* 6, 3: 101.
Stronge, J.H. (1992) "Programs with promise: educational service delivery to homeless children and youth," in Stronge, J.H. (ed.) *Educating Homeless Children and Adolescents: Evaluating policy and practice*, Newbury Park, Calif.: Sage.
Suchman, D. (1990) *Public/Private Housing Partnerships*, Washington, D.C.: The Urban Land Institute.
Sullivan, W.P. (1992) "Reclaiming the community: the strengths perspective and deinstitutionalization," *Social Work* 37, 3: 204–209.
Sykes, R. (1994) "Older women and housing – prospects for the 1990s," in Gilroy, R. and Woods, R. (eds) *Housing Women*, London: Routledge, 75–100.
Taylor, C. (1995) *Philosophical Arguments*, Cambridge, Mass.: Harvard University Press.
Taylor, G. (1984) *A Study of Hostels for Young People in London*, London: Greater London Council.
Ternowetsky, G.W. and Thorn, J. (1990) "Nowhere to go but down," *Perception* 14, 4: 36–39.
Thompson, L. (1988) *An Act of Compromise: An appraisal of the effects of the Housing (Homeless Persons) Act of 1977 – ten years on*, London: SHAC/Shelter.
Timmer, D.A., Eitzen, D.S., and Talley, K.D. (1994) *Paths to Homelessness: Extreme poverty and the urban housing crisis*, Boulder, Colo.: Westview Press.
Titmuss, R.M. (1974) *Social Policy: An introduction*, London: George Allen and Unwin.
Tobin, G. A. (ed.) (1985) *Social Planning and Human Service Delivery in the Voluntary Sector*, Westport, Conn.: Greenwood Press.
Todd, E. (1991) *The Making of Modern France: Ideology, politics and culture*, Oxford: Basil Blackwell.
Toffler, A. (1984) 'Foreword,' in Prigogine, I. and Stengers, I. *Order out of Chaos: Man's new dialogue with nature*. Toronto: Bantam Books.
Toth, J. (1993) *Mole People: Life in the tunnels beneath New York City*, Chicago: Chicago Review Press.
Townsend, P. (1993) *The International Analysis of Poverty*, New York and London: Harvester Wheatsheaf.
Townsend, P., Simpson, D., and Tibbs, N. (1985) "Inequalities in health in the City of Bristol: a preliminary review of statistical evidence," *International Journal of Health Services* 15, 4: 637–663.
Turpijn, W. (1988) "Shadow-housing: self-help of dwellers in the Netherlands," in Friedrichs, J. (ed.) *Affordable Housing and the Homeless*, Berlin: Walter de Gruyter, 103–113.
U.S. Bureau of the Census (1951) *Census of Population and Housing*, Washington, D.C.: U.S. Department of Commerce, Bureau of the Census.
—— (1982) *Annual Survey of Housing*, Washington, D.C.: U.S. Department of Commerce, Bureau of the Census.

—— (1989) *A Status Report on Hunger and Homelessness in America's Cities, 1988*, Washington, D.C.: U.S. Bureau of the Census.

—— (1990a) *Statistical Abstract of the United States*, 110th edition, Washington, D.C.: USGPO.

—— (1990b) *Current Population Reports*, series P-60, Washington, D.C.: USGPO.

—— (1991) *Statistical Abstract of the United States*, 111th edition, Washington, D.C.: USGPO.

U.S. Conference of Mayors (1987) *The Continuing Growth of Hunger, Homelessness and Poverty in America's Cities: 1987*, Washington, D.C.: U.S. Conference of Mayors.

—— (1989) *A Status Report on Hunger and Homelessness in America's Cities: 1988*, Washington, D.C.: U.S. Conference of Mayors.

—— (1995) *A Status Report on Hunger and Homelessness in America's Cities: 1994*, Washington, D.C.: U.S. Conference of Mayors.

U.S. Congress (1930) House Judiciary Committee *Unemployment in the United States: Hearings*, 71 Cong., 2nd Session.

—— (1983) House Committee on Banking, Finance and Urban Affairs *Homelessness in America*, Washington, D.C.: USGPO.

—— (1986) House Committee on Veterans' Affairs *Homeless and Unemployed Veterans*, Washington, D.C.: USGPO.

—— (1987) *Public Law 100-77: Stewart B. McKinney Homeless Assistance Act*, Washington, D.C.: USGPO.

—— (1990) House Committee on Energy and Commerce, Subcommittee on Health and the Environment "Health care for the homeless: hearing before the subcommittee," June 15, Washington, D.C.: USGPO.

U.S. Department of Housing and Urban Development (1984) *A Report to the Secretary on the Homeless and Emergency Shelters*, Washington, D.C.: Office of Policy Development and Research.

—— (1989) *A Report on the 1988 National Survey of Shelters for the Homeless*, Washington, D.C.: Office of Policy Development and Research, Division of Policy Studies.

—— (1995) *Review of Stewart B. McKinney Homeless Programs Administered by HUD: Report to Congress*, Washington, D.C.: Office of Policy Development and Research.

U.S. Federal Public Housing Authority (1945) "Minimum Physical Standards and Criteria for the Planning and Design of FPHA-aided Low-Rent Housing," Washington, D.C.: National Archives and Records Service, RG 196.

U.S. General Accounting Office (1985) *Homelessness – A Complex Problem and Government's Response*, Washington, D.C.: USGPO.

U.S. House of Representatives (1983) Committee on Banking, Finance and Urban Affairs *Homelessness in America*, Washington D.C.: USGPO.

U.S. House of Representatives (1984) Subcommittee on Housing and Community Development *Homelessness in America II*, Washington, D.C.: USGPO.

U.S. Senate (1993) Committee on the Judiciary, Subcommittee on Juvenile Justice *The State of Youth at Risk and the Juvenile Justice System: Prevention and intervention*, Washington, D.C.: USGPO.

Viden, S. (1985) "Approaches to the improvement of social/public housing – national background – Sweden," in *Approaches to the Improvement of Social/Public Housing*, The Hague: IFHP.

Wade, J. (1986) "Wartime Housing Limited, 1941–1947: Canadian housing policy at the crossroads," *Urban History Review* 15, 1: 41–59.

Wagner, D. (1993) *Checkerboard Square: Culture and resistance in a homeless community*, Boulder, Colo.: Westview Press.

Waldie, P. (1991) "Food banks having to ration," *The Globe and Mail*, October 21: A4.

Warner, S.B., Jr. (1962) "Preface" in Woods, R.A. and Kennedy, A.J. *The Zone of Emergence: Observation of the lower middle and upper working-class communities of Boston, 1905–1914*, Cambridge, Mass.: MIT Press.

Watson, S. and Austerberry, H. (1986) *Housing and Homelessness: A feminist perspective*, London: Routledge and Kegan Paul.

Weir, M., Orloff, A.S., and Skocpol, T. (1988) *The Politics of Social Policy in the United States*. Princeton, NJ: Princeton University Press.

Weisbrod, B.A. (1989) "Alternative to mental hospital treatment: II. Economic benefit-cost analysis," *Archives of General Psychiatry* 37, 4: 400–405.

Weller, M. (1991) "The need for asylum," in Page, M. and Powell, R. (eds) *Homelessness and Mental Illness: The dark side of community care*, London: Concern Publications, 41–46.

West, D.J. (1977) "The chronic offender," *Hospital Medicine* October: 48–50.

Whitehead, M. (1987) *The Health Divide: Inequalities in health in the 1980s*, London: Health Education Authority.

Williams, F. (1993) "Gender, 'race' and class in British welfare policy," in Cochrane, A. and Clarke, J. (eds) *Comparing Welfare States: Britain in international context*, London: Sage, 77–104.

Williams, P. (1991) *Alchemy of Race and Rights*, Cambridge, Mass.: Harvard University Press.

Williams, R.H. (1986) *Learning from Other Countries: The cross-nation dimension in urban policy-making*, Norwich, U.K.: Geo Books.

Wolch, J. (1990) *The Shadow State: Government and voluntary sector in transition*, New York: The Foundation Center.

Wolch, J.R. and Akita, A. (1989) "The federal response to homelessness and its implications for American cities," *Urban Geography* 10, 1: 62–85.

Wolch, J.R. and Dear, M.J. (1993) *Malign Neglect: Homelessness in an American city*, San Francisco: Jossey-Bass.

Wood, E.E. (1923) *Housing Progress in Western Europe*, New York: E.P. Dutton.

Woods, R.A. (ed.) (1903) *Americans in Process: A settlement study*, Boston: Houghton-Mifflin.

Woolley, T. (1988) "Ten topics for participators," *Open House International* 13, (3): 49–53.

World Food Day Association (1993) *World Food Update*, Ottawa.

Wright, J. D. (1988–89) "Forum," *Issues in Science and Technology* (Winter).

—— (1989) *Address Unknown: The homeless in America*, New York: Aldine de Gruyter.

Wright, J.D. and Weber, E. (1987) *Homelessness and Health*, Washington, D.C.: McGraw-Hill.

Yanetta, A. and Aldridge, R. (eds) (1992) *Housing, Homelessness and Europe*, a report based on papers given at a conference in Edinburgh, Edinburgh: Institute of Housing in Scotland and Scottish Council for Single Homelessness.

York, G. (1992) "Study undermines belief in 'lazy welfare bum'," *Globe and Mail* (Toronto), July 4.

—— (1992) "Defending the definition of poverty," *Globe and Mail* (Toronto), July 16.

York, M.L. (1991) "The Orlando affordable housing demonstration project," in *Journal of the American Planning Association* 57, (4): 490–493.

INDEX